Copyright 2017 by David W. Braeutigam

All rights reserved. No part of this book may be reproduced or transmitted in any form or by any means, electronic or mechanical, including photocopying, recording, or by an information storage or retrieval system without written permission from the author, except for the inclusion of brief quotations in a review.

The book contains anecdotes from the author and many told to the author. The author does not claim to the accuracy of the stories and has tried to leave out the names of the hospitals. Maybe I should have named the book Tall Tales of the Biomed and it would be more realistic.

This book also contains some awesome pranks which are not recommended to be used in your workspace today due to concerns from HR and probably your fellow biomed.

Why are you reading this section of the book? No one reads the copyright information. This is boring – go read the funny stories.

Tales of the Biomed

A collection of funny anecdotes, serious stories and photos from the biomedical equipment technicians and clinical engineers across the world as told to the author.

By David W. Braeutigam

MBA, CHTM, CBET, ITIL

Dedication

This book is dedicated to the biomedical equipment technicians[1] and clinical engineers[2] that service and support medical equipment across the world. With their dedicated service the medical equipment is safe to use for our patients and our staff in our hospitals and healthcare systems everywhere.

Acknowledgments

I would like to thank the following people for providing the photos, serious stories and funny anecdotes that made this book possible. David Bilder, Richard Swim, Steve Juett, Jeff Butler, Richard Roa, Don Hildebrand, Tim Dess, Daniel Irving, Tim Huffman, Doug Dreps, Travis Kobernick, Travis Recksiek, Paul Stodolka, Larry Hertzler, Andrea Brainard, Matt Baretich, Adam Fiske, David Mason, Justin Donovan, Craig Westbrook, A Zahi Adl, Chace Torres, Luis Jurado, Joe McClain, Thomas Putt, Donny Letson, Dave Reed, Mindy Chen, Jay McClure, Mike Kerchner, Corey Miller, Justin Wallace, Shannon Ouimet-Amaro, Matt Burns, Cody Brown, Pete Martin, Chris Braaten, Reginald Jones, Mitch Smart, Jeremy Hendrick, Richard Eliason, my brother Joe Braeutigam and of course my wife Tawnya, for putting up with my stories and letting me take away from our evenings together to work on this book.

Table of Contents

Introduction 1
Practical Jokes and Pranks... 7
Adventures of the Road Warrior 21
Badges, I Don't Need No Stickin' Badges! 25
Technology and the Biomed.. 31
Unusual Service Calls.. 38
Sticky Notes and Nurse's Tape 58
Biomeds are a Resourceful Bunch 83
Fitzsimons Stories... 88
Other books by the author.. 97
References... 98

Introduction

The field of Biomedical Engineering, Clinical Engineering or Healthcare Technology Management (as it is called today) has been in existence in the civilian world since the late 1960's or early 1970's. For the U.S. Military it has been around since the 1940's and was called Medical Maintenance. Thousands of individuals have called this field their career and have great stories to tell. This book is a collection of some of those stories.

I was introduced into the field as a junior at Eastwood High School in El Paso, Texas way back in 1975. One of the tests we were required to take was the ASVAB test – the Armed Services Vocational Aptitude Battery test. I guess it helped you (or your career counselor) decide what you were best set out to be in your career. Apparently I scored well in math and electronics and they suggested I join the U.S. Army as a medical maintenance technician. This sounded like a great idea to me. Little did I know this is what I would do for the rest of my career. But I think I was destined to be in the biomed field – take a look at this photo of my brother in traction with me there to provide technical support! Biomedical Engineering was in my DNA.

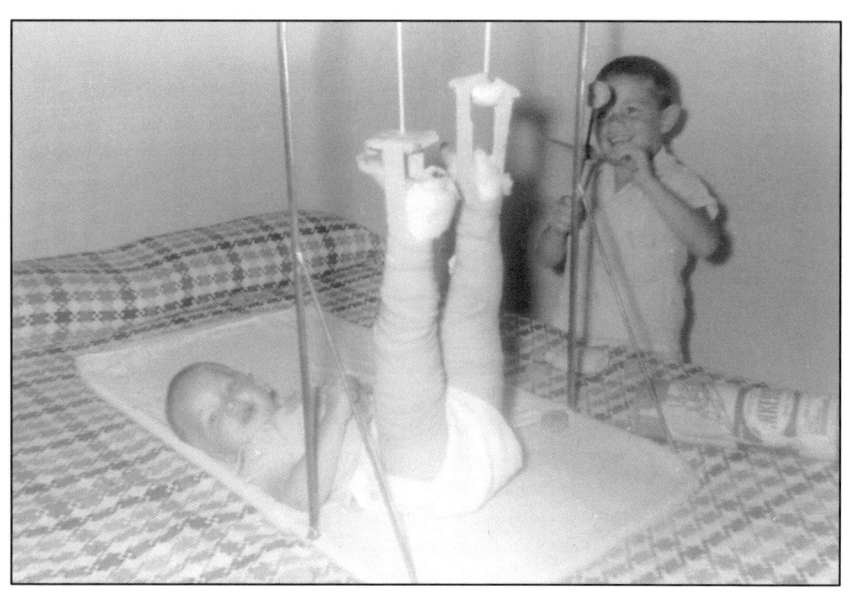

David Braeutigam providing technical support to his little brother Joe in traction circa 1963 [3]

A short time later I decided that was the best thing for me so I joined the U.S. Army. I would later tell my mother about joining the Army – another story for a different time. So there I was with other 17-18 year old guys standing nude in some sort of military clinic getting examined and prodded by some doctor. Apparently I passed and I was given a deferred enlistment. I officially joined the military around March of 1976 with a reporting date of August 1976.

I graduated from Eastwood High School in May of 1976 and enjoyed the summer before reporting to Fort Bliss Army

Base in my hometown of El Paso, Texas. No fancy flights or troop transports for me. My mother took me to the post in our 1963 baby blue Chevy Impala and dropped me off for basic training.

David Braeutigam and his brother Joe in 1976 with typical 1970's hair styles – pre Army [4]

One of the first things they do is to make sure you meet all the requirements of a military man. These means you get a haircut. A point of reference for the reader. This was 1976. I had an afro so it was like shearing a sheep for the barber as I got my hair cut to meet the military regulations. Then you were sent to the 'tailor' to get custom fitted green clothing and nice combat boots. You are in the Army now!

David Braeutigam U.S. Army September 1976 [5]

Eventually I would graduate in eight weeks from basic training at Fort Bliss and then I was sent to the medical maintenance school in Aurora, Colorado. Since the Army is well known for their efficiency they sent me to San Antonio and Fort Sam Houston. Why you ask? That is because all the training for the medical field was done at Fort Sam Houston back then. Of course when I arrived in San Antonio they asked why I was there and eventually they got me on another plane to Aurora, Colorado.

I would eventually graduate from the United States Army Medical Equipment and Optical School (USAMEOS) in March 1977 and was sent to Walter Reed Army Medical Center in Washington DC. I stayed there for one year and was sent back for advanced training at USAMEOS and graduated in December 1978. I was then sent to the 32nd Medical Supply Optical and Maintenance (MEDSOM) at Fort Bragg, North Carolina. Fort Bragg was unlike Walter Reed and Fitzsimons. It was the REAL Army. You wore starched fatigues, saluted officers, ran for physical training (PT) and basically acted like a real soldier.

When my military tour was up I moved back home to El Paso, Texas to work at a local hospital and decided to go to the University of Texas at El Paso (UTEP) for an Electrical Engineering degree using the GI Bill. Things changed and I moved to Dallas to live with my brother and I eventually found a job at a local hospital as a BMET. That's the story of my early life so let's move on to the biomed stories!

David Braeutigam U.S. Army BMET in 1977 [6]

Practical Jokes and Pranks
Conductive Shoe Covers

It was common in the 'old days' to pull practical jokes on each other. One of the first ones I remember had to do with the monthly testing of equipment in the operating rooms. This was when I was stationed at Walter Reed Army Medical Center in Washington, D.C. in 1977. Back in those days we still used flammable anesthetics so you had to have conductive floor testing in the operating room. This also meant you had to have booties for your shoes with conductive straps that were tucked into your socks. All of the medical equipment in the operating room had chains that drug on the floor to keep everything grounded to the conductive grid in the floor. One evening we are putting on our scrubs and I am helping a newbie put on his scrubs and conductive shoe covers. As we are putting them on I asked him if he had a right and left bootie. He looked at me and said he was not sure. I kindly told him I would help him locate the left and right booties since they were very hard to distinguish. We spend several minutes looking for a left and right booties and I finally had to tell him that we will just have to wear two right booties. I am not sure if he ever found out there is NO left and right booties.

The Balancing Cart

Tim Huffman has lots of stories around harmless pranks in the biomed world, but the one that left a shop in tears is the one Tim will tell here.

We had a new guy in the shop. Gene. He wasn't sure exactly how he was going to fit in and you could see the nervousness but also excitement in joining our team. We had around 8 people in this shop.

We had an idea that putting a tool cart on top of someone's desk would be funny. The tech couldn't get the cart down without help, and obviously they couldn't have their cart on their desk long so this would take an immediate action from the recipient of this prank.

So, one specific tech that was not in the shop at the time that a few of us dreamed-up this trick was…. Well let's say Sarah. Sarah's cart was there and available so a few of us including Gene and myself cleared off her desk and lifted her cart onto her desk. We tried for several minutes to get the cart to balance itself on her desk but no matter how we manipulated it the cart would not stay on its own. The door to the shop starts to open and of the three of us that lifted the cart originally, two of us let go leaving Gene the new guy holding the cart in place on top of her desk.

Sure enough Sarah was the one entering the shop and as the rest of us tried so very hard not to laugh she came across Gene, which she has no familiarity with, holding her cart on top of her desk. At this point the rest of the shop is crying we are laughing so hard as Sarah asks Gene "so… whatcha doing there Gene?"

Gene looked at the rest of us and expressed some choice words that I won't repeat here. I can say that he held the cart there as he looked at his new shop mate for at least a few minutes as the rest of us dried our laughing tears. Eventually we helped him bring the cart back down to the floor but that is a moment I will never forget.

The Aqua Fuse

Inexperience makes us do a lot of unusual things. When I was stationed at Fort Bragg I was in a field unit. You rarely worked on medical equipment - instead you played Army all day to get prepared to go to a war. This was the beginning for me to decide to get out of the military and join the civilian ranks. Hey, maybe I could grow my Afro back. Anyway, I digress. For several months I was fortunate to actually work in the medical center at Fort Bragg. Back then we had Spacelabs monitoring for the ICU. We had a defective monitor and my job was to determine the problem. As I took the monitor apart it had a massive main board and I determined the fuse was defective. This was not a fuse in a socket but a fuse with pigtails that was soldered to the main board. I unsoldered the fuse and went looking around the shop for a replacement. I could not find another pigtail fuse with the correct rating. I wondered if I could unsolder the fuse from the pigtail holder, replaced it with the correctly rated fuse and get the unit back to the ICU properly working.

Don't get ahead of me here – I am still young and naïve and didn't realize that soldering on a fuse will melt the internal fuse link. I finally came to the conclusion that this was not going to work. But I also was the first inventor of the Aqua Fuse! You may have never heard of the Aqua Fuse so let me tell you how it works. You take a pigtail fuse that some new guy has been soldering on for awhile and clean out the internal fuse link. You refill the empty space with the liquid used to clean the surface of your work space and viola you have an Aqua Fuse.

That is the easy part – now you have to have a story for the other guys in the shop that wonder what you are doing. Of course another biomed tech comes by and asks what you are doing. You tell him you are replacing the Aqua Fuse in this Spacelabs monitor. He astonishing asks 'What is an Aqua Fuse?" Of course I had to explain to this guy that the Aqua

Fuse is used to fuse the circuit on this main board but at the same time it helps you balance the main board (using the bubble in the fuse) so all the electrons flow evenly around the circuit. It was a genius invention but I never got around to a patent[7]. I told him that the different colors of fluid (green, blue, orange, etc.) would be the indicators for the fuse rating.

The morgue and the newbie

Joe McClain shared this story with me. As you may know most maintenance shops are normally in the basement and of course so is the morgue. I will not tell you the base that we did this because even though it was years ago we don't want anyone to get in trouble. We had a buddy who was the morgue tech and he would let us know when they had a cadaver on the table and call in a repair on the Surgical Light over the table. We would disconnect the power lead in the light under the control panel. We would make sure lights were out and then send in the newest, youngest tech we had with the work order to repair it. We had a hidden camera in the room so we could record the entire repair. We also had a remote CRT so we could see and hear what was going on in real time. You can imagine the reaction of the kid when he first turned on the light to the room. During his repair we would transmit low sounds into the room like soft blowing wind that might come out of a body. The kid was scared and nervous during the entire repair and when we played the video back for the shop afterwards, the laughter was loud and boisterous.

Mother-in-Laws

I was listening to the radio one day and the DJ mentioned he always sent his mother flowers for Mother's Day. The other DJ joked that "Yes but you send them COD!" I thought what a great idea! So I called the local florist in the hometown of my mother-in-law, Lela Fanning. The florist knew Lela so I told them to send a dozen roses to her at the fancy women's store where she worked. I would pay for them but I wanted them to tell her they were COD! They laughed and agreed. Needless to say I got a quick call from my mother-in-law about this and I still hear about it today. I have a great relationship with my mother-in-law so don't try this unless you do also. She pranked me a couple of times after this and we had to call a truce. Maybe it is time for another one …

Pink Socks and a Laurel and Hardy Handshake

I was fortunate to be the speaker at the Association for Advancement of Medical Instrumentation (AAMI) Conference in Tampa, Florida, back in 2010. My job was to introduce the main speaker, which I believe was someone from GE Healthcare. The AAMI representative, Chris Dinegar, wrote my speech so it was pretty simple to do. I had worn a light grey suit with pink socks and a pink tie. I thought it was pretty appropriate tropical attire for Florida.

David Braeutigam speaking at AAMI Conference in Tampa 2010 [8]

I had always wanted to do the Blazing Saddles act of introducing someone with a "Laurel and Hardy handshake"[9] but Chris was not going to have anything to do with that. He was very worried. I told the GE speaker of my plan and he thought it was funny also, but I decided not to do it. I introduced the GE speaker like a good 'soldier' and went back to my chair. At the end of the GE speaker's speech I was called back to the stage to welcome everyone to the conference and to tell them to have a great time at the meetings. Since it was Florida, I stepped away from the podium and pulled up my pants legs to reveal my pink socks - which I thought went along with the atmosphere of the meeting. The problem was I was not next to the microphone and no one could hear me so everyone thought I had NO socks on. Hey, that worked too!

We're working down here!

Here is another one from Joe McClain. We had a secretary that was the Commanders that was always giving the CBET's or BMET's a hard time and any chance she would get, she would blame, or try to make them, look bad. Why; I don't know. At any rate the latrine upstairs was under renovation and she had to use the one across from our shop. Jessey, one of our senior technicians, rigged a small intercom speaker under the commode. In those days nothing was wireless so he had to run a speaker wire through the ceiling from the maintenance shop to the latrine. He asked us to keep an eye out and let him know when she went in, which of course we did. When he got the nod he waited about 2 or 3 minutes and then he transmitted over the intercom and said, "CAREFUL LADY! WE'RE WORKING DOWN HERE".

Within 10 to 12 seconds she came running out of the bathroom red as a beet straightening her dress. Jessey removed his toys from the scene in less than 3 minutes not leaving a clue that he had ever been there. Just as he finished up, she and the Commander came down like storm troopers and wanted to know if we had seen anyone or anything unusual in or around the ladies bathroom. Of course, all of us knew nothing and Jessey came away without anyone detecting his deed. For some reason after that she left the BMET's alone and never bothered them again. She also refused to use that bathroom for as long as I was stationed on that base. It was actually funnier than hell. I'll never forget that look on her face when she came out of that bathroom. I wish we had a snapshot of her exit.

Super Glue and tools

I had a good prank pulled on me just shortly after I started working at a large medical center in Dallas, Texas. I made the mistake of having a girlfriend come by for lunch and meet the guys. We met all the guys at a local barbeque restaurant and everyone was very nice but sure seemed to be in a hurry to get back to the shop. They told me not to worry as they hurriedly ate their lunch and were ready to leave. So my girlfriend and I stayed and enjoyed our lunch. After lunch I took her back to the shop to see my work area. Now I could see why the guys were in such a hurry. They decided to glue down all my tools and supplies to my desk. They also poured packing peanuts and trash all over the desk. I am sure she was very impressed! This time the prank was on me!

Roaches and scientific experimentation

Having lots of new tools and supplies at your disposal also makes your mind wander. It was not uncommon to find a cockroach or two in our shop during those early days. It was also not uncommon to have freeze spray in the shop. We used freeze spray to spray on electronic components to see if problems were caused by a circuit overheating. It could help us pinpoint the problem area when the problem disappeared. It also caused lots of excitement in the shop spraying each other with freeze spray. Remember the field is still very young and most of us are still in our 20's. Anyway back to the cockroach problem. I wondered (from a scientific perspective only) if freeze spray could kill a cockroach. Luckily we had a cockroach available for testing and I applied the freeze spray. He turned sold white and did not move. A short time later the cockroach came back to life. Amazing! Now I am wondering what other tests we can perform on him. Next we put him in a centrifuge and spun him at several thousand RPMs. With the test concluded we removed our test subject and found he lost some legs but was still alive.

Next test was the microwave oven. This was no test and he met his demise. For some reason the girls in the shop were not too happy that we used the microwave for this test. They just don't appreciate scientific research!

Open house and beer cans

One year my boss, Richard Roa, was leaving our hospital to go work for a major medical manufacturer. We had a big party at his house in north Dallas. Somehow I got the idea to start hiding the empty beer cans in his dog's food, in his toilet, on top of his kitchen cabinets, etc. Others joined in the fun. Years later this would backfire on me as it had since become a tradition to hide beer cans at people's homes when they had a party. My wife, Tawnya, and I had bought our first house and held a house warming party. Sure enough people started hiding beer cans – some right in front of my wife and in front of me. I understood the 'tradition' but my wife was furious. It took us weeks to find all the beer cans. She is still mad about that today. Just ask her!

Another thing we did for Richard Roa just before his last day was to put a biomed control number on his car. Back then cars had chrome bumpers so the control number was able to stay attached. I think after a year, we contacted him and said his Toyota was due for a PM! I wonder if that asset is still in the database?

Sterilizer and the tattooed guy

When we used to do service for non hospital owned clinics and had a guy bring a table top sterilizer into our office for repair. This was a large guy with tattoos everywhere. He said he just bought a used car and that the sterilizer was in it and he used it at his tattoo shop. OK, that's a good thing but this guy still looked shady. This was back in the late 1980's and tattoos were not as prevalent as today.

So I fixed the table top sterilizer and called him to come pick it up. He said he didn't have the money at the current time but would call when he had it.did.

This went on for months and we finally starting using the sterilizer as a loaner when we were repairing other ones. Finally out of nowhere the guys walks into our office ready to pick up the sterilizer. Our secretary knew what we were doing and she was scared to death. She called me to the front. Thinking quickly I said it had been awhile since we had seen him and we had it offsite in storage (well it was offsite). I said it will take a day or two to pick up. He was OK with that and came back later to pick up his sterilizer.

This smell horrible – take a whiff!

Remember the old Gomco Thoracic Drainage Pumps from the early 1980's? They had a heater inside the unit that created a very slight suction just enough to drain the thoracic cavity. If you did not realize this you would think it was broken.

Anyway, I had one that was not creating suction at all. I took the unit apart and the heater was shorted by the thoracic fluid sucked into the chamber. The smell was horrible. But you are a biomed so you know what to do – pass it around for everyone else to smell. We all gagged at the smell. It finally got to the time to pass it around to the female office personnel and they were NOT happy at what I requested of them.

Blood is everywhere …

We used to have an office person named Barbara that worked the front desk. She was a great lady but I loved to pick on her and pull jokes and pranks on her. I must have eaten out that day and I had several packets of catsup on my desk. The light bulb goes off in my head.

I took a hammer and slammed it down on the desk, screamed loudly and then put catsup all over my wrist. I then ran to the front to show her what happened. I thought she was going to faint so I finally had to tell her it was catsup. I think she hit me.

Adventures of the Road Warrior
Ultrasound Gel Missile

Those road warrior biomeds that service multiple sites have a different set of tales to share than those in a single hospital. I was once driving our company van loaded with test gear and medical equipment between two of our hospitals in North Texas when all of a sudden something shot across the van and bounced off the front windshield. I tried to keep my eye on the road but at the same time, trying to figure out what the heck had just happened. Did someone just shoot at me? I was driving in downtown Dallas, so that was a possibility. Did something just break on the van? I checked the dashboard and everything seemed to be OK. Once I finally had a chance to park the van and investigate what hit the front windshield. My crack detective skills found the top of an ultrasound gel bottle that must have come dislodged and shot across the van. I finally determined I must have made an abrupt lane change that shifted the therapeutic ultrasound in the back of the van and pinched the ultrasound jell container which forced the cap to shoot across the van. I was able to determine this because there was a large amount of ultrasound jell all over the back of the van. I now realized why so many service vans have the wire cage between the front seats and the back of the van. It is there to protect the driver from equipment entering the driving area!

No ties please!

In the 'old' days the biomed tech wore a dress shirt, a tie and dress slacks as their work attire. One particular day I was working on a very large front end processor (FEP) that connected downtown to a mainframe (this was around 1988). We had an error and I troubleshot it down to a particular FEP. I was all alone in the computer room and decided to open the chassis and inspect this issue. As I did, my tie got caught in the power supply fan! I am now stuck in the power supply chassis with no one around and no way to call for help. I thought about the situation and finally reached around front and turned off the power and slowly pulled my tie out of the power supply fan. I learned a valuable lesson about wearing a tie and working on electronic equipment – tuck you tie into your shirt before proceeding!

Receipts

Back when I did field service for our outside hospitals and clinics it was common to have to pay for parking. The problem was most did not give receipts. Even though most involved just a couple of dollars it could added up over time.

Since we had to have a receipt for all expenses I needed to find a way to fix this. We had just gotten laser printers to go along with our Mac computers. The Mac came with MacPaint which was a drawing program. I designed a typical receipt for parking so I could get reimbursed. I told my boss, Richard Roa, about it and he said "I wish you would not do this." Ironically he came to me a short time later and asked "Can you make some receipts for me?"

Another time, I was at a class in California for a computer system, and I did not have receipts for all my meals. Back then I would normally just eat at McDonald's so I decided to duplicate the receipts for my meals that I did not have. At home, I had a Commodore 64[10] and a Commodore 1520 Printer Plotter.[11] I basically had to write a BASIC program to duplicate the receipts. They turned out great as I even duplicated the black and red ink needed to exactly duplicate the receipts.

Wanna have a good time?

When I did field service we had a van assigned to each of us. We also took the van home every day. One morning I took my normal trip to McDonald's for breakfast. It was a cold day as I got my order of a sausage biscuit, hash browns and a diet coke through the drive-thru. I stopped after getting my order to put some catsup on the sausage biscuit, when I noticed a scantily clad young lady come up to the van. I rolled the window down to see what she wanted. She wanted a ride just up the street. I wouldn't normally do this but it was COLD outside and she didn't have a lot of clothes on. I decided not to judge and gave her a ride. Before long she was propositioning me and continued to want to go further. Finally I let her off at a hotel. I'll never do that again!

Badges, I Don't Need No Stickin' Badges!
Conference Badges

Conferences are a lot like the military. The more ribbons you have, the more important you must be. The goal of a conference attendee is to have the most badges and thus, they must be the most important person in attendance. I think it was during the AAMI Conference in 2006, I was going to the last presentation of the day on the final day of the conference. As I was walking to the meeting room, I happened upon some of the AAMI booths. They were still there in case somebody had to check in. I figured, why not stop at the booth and see if I can procure some extra ribbons. It is the last hour of the last day of the conference, so who would be concerned. So I stopped at the booth and ask the woman if I could have some extra ribbons. She said sure – which ones would I like. I stated "All of them!" And so I got every ribbon they had. We stuck all the ribbons together on my conference name badge and it reached halfway to the floor!

David Braeutigam with *every* AAMI badge. AAMI Conference 2008 [12]

Based on my prior logic, I must now be the most important person at the conference. I was more important than Mary Logan (then President of AAMI), more important than Steve Campbell (COO of AAMI) and definitely more important than Bob Stiefel! As I started toward the meeting, room panic set in. What if someone from AAMI took offense to what I just did and the woman at the booth got in trouble? I hastily hid the badges under by sports coat and walked awkwardly to the conference room as I tried to keep the long set of ribbons from falling out. Along the way, I ran across several people I knew and I opened my sports jacket to reveal my new collection of ribbons. We all had a good laugh, but I don't think I ever found Bob Stiefel to 'flash' him my set of ribbons!

Binseng Wang and duct tape

Binseng Wang is very well known in the biomed community. He has authored numerous articles, worked on lots of AAMI committees and spoken about our field at countless conferences. He is NOT known for the lack of speaking up to give his opinion. However, at the AAMI 2008 conference he had a trick up his sleeve. I believe I was the moderator for a talk by George Mills from The Joint Commission, when we got to the portion for questions. Binseng Wang approached the stage with duct tape covering his mouth. It was hilarious because we knew he wanted to speak his opinion on the latest Joint Commission changes for biomed! I was fortunate to get the photograph and now you can see that sometimes Binseng CAN be speechless!

Binseng Wang with duct tape over his mouth. AAMI Conference 2008 [13]

Two Kings and their Crowns

I was fortunate to be selected the AAMI/GE Healthcare BMET of the Year in 2008 and my boss at the time, Ken Maddock, was awarded the AAMI Leadership Award for his career achievements in the field. It was quite an honor for both of us and also an honor for our healthcare system since we worked together. I decided to poke some fun at us and bought some crowns from Burger King on the way home from the conference. I asked our boss (the CIO) to present the crowns at our managers' meeting. Don't you think we look regal?

Two Kings and their crowns [14]

Don't mess with my name badge

Most biomed shops have interns in their shop. In the early 1980's we used to get biomedical engineering students from Texas A&M University and other university programs. They may be students, but they are not immune to practical jokes.

One year we had an intern, I think here name was Rhonda, and her name badge said 'Clinical Engineering Asst.' She wore it proudly on her white lab coat like we all did back then. I immediately saw an opportunity. What if I took Wite-Out[15] and blotted out the 't' and the '.' from Asst.? The prank was on. She proudly wore her name tag throughout the hospital until someone finally pointed out the error in the spelling. She was furious and for some reason came directly to me and accused ME of doing it. I, of course denied it – until today. Still one of my best all-time pranks in my opinion. I am sure she feels otherwise.

Technology and the Biomed
X-10 remote control

In the early 1980's while working as a BMET at a large medical center in Dallas, I decided to play a trick on a fellow Clinical Engineer Rex Moses. Rex usually arrived at work around 8:15 every morning, I had some time one morning to set up the trick. I am a geek at heart probably like a lot of biomeds, I just started to use X-10[16] to automate my apartment I was living in at the time. I brought a lamp module and a remote control for the lamp module from home to set up the trick. The remote control allows you to remotely turn on and off the light from across the room. Rex sat two cubicles away from me so I could easily monitor his reaction as I turned on and off his desk lamp. After Rex arrived I started to remotely turn off his light and wait for his reaction. I waited a short time and turned the light back on. Rex started to wonder why the light was malfunctioning. He unscrewed the bulb to see if anything was wrong. The bulb seemed to be OK. I waited for a short time, trying to hide my amusement, and cycled the lamp on and off again. After a short time everyone in the shop was aware of what was going on and we had to let Rex know about the trick. We all had a good laugh!

Vibrating Pagers

Probably around the middle 1980's is when the first vibrating pagers came out. Every biomed I knew carried a pager so we knew there had to be a trick we could pull on each other with the vibrating pagers. We determined the best time to try this trick was when another biomed in the shop was deep into a circuit repair, page the biomed, and have his pager go off. You can imagine his surprise when he thought he was getting shocked by the circuit he was working on because of the vibrating pager!

Scotty, Beam me up!

The early pagers we used in biomed were voice pagers. You would call a number and record your voice message. The message was sent to the pager and your voice was repeated out loud on the pager. I was ready to go to lunch one day and I wanted to see if Richard Swim wanted to join me. I figured I would page him with a message. But I didn't want to leave a boring message about going to lunch so I paged him with the message of "Scotty, beam me up!" with a slight Irish accent. Little did I know Richard was in a pretty high level meeting and the message backfired on me! Richard was not happy about being embarrassed by me in his high level meeting!

Pagers and Drug Dealers

We all wore pagers in the early days of biomed and probably some still do today. In the early days, only doctors and drug dealers wore pagers. Paul Porter was working for us and had two pagers – one for normal work and one for on-call. He went to a family reunion one weekend and everyone thought he must be a drug dealer to be carrying not one, but TWO pagers on his hip. I am pretty sure Paul wasn't dealing drugs – at least not that I was aware of!

The VIC-20[17]

At one hospital I worked at, we were having theft issues with batteries missing from our shop. We assumed someone from Engineering or someone with a master key was stealing the batteries at night. I happened to have a Commodore Vic-20 and decided to write a program to determine when the door to our shop was opening. It was a simple Basic program with an interface to a magnetic switch to the front door connected to the joy stick port of the computer. Remember, this was around 1984 so we didn't have the luxury of cameras to catch the thief. My program did determine someone was coming in after hours, but that is all it was able to do.

We had the computer locked in our director's office and I later wrote a program to have it 'ding' every minute. He did not think it was funny and I eventually had to turn it off.

Hacking an Air Force Computer

Around 1982 I convinced my boss to let me go to the University of Texas at Arlington so I could get my Electrical Engineering degree. I would go full time and work almost full time at the hospital. This lasted only one semester. But I got a chance to pull a great prank.

I had just left my last class, probably calculus, and was on my way to drive to work. This would be around lunch time. We were walking across campus when we came upon an Air Force Recruiting trailer. We stepped inside and it was full of posters and recruiting information and I spotted a computer. It would prompt you to type in your name and would then put information on the screen with your name. A pretty basic program (no pun intended). I thought – I wonder if I could 'hack' this? I knew a little BASIC programming, so I hit some keys and the program stopped. It was a Radio Shack computer which I had never worked on but I was not concerned. I listed the program while at the same time, looking out the corner of my eye for the Air Force recruiter to come through the door, anytime. I found the listing in the program where it assigned the name (probably NAME$). I inserted a line and typed something like 201 LET NAME$=DICKHEAD. I hit return and ran the program. I typed in my name and the name returned with DICKHEAD. Perfect! I now hacked a government computer as a prank. We waited what seemed like 30 minutes, and no one came into the trailer to try the new program! I finally told my friend I had to leave to go to work. I wish I was there to see what happened when someone typed their name in. I would have loved to see their face and the face of the Air Force recruiter!

Environmental Control Unit

In the late 1980's I was fortunate to work with some very talented Clinical Engineers and Biomed Techs at a hospital in Dallas. In those days if you needed medical equipment and it wasn't available, we would just design it. I supported a small rehab hospital and we had a quadriplegic at the hospital. The staff contacted me and a Clinical Engineer to see if we could design something so the patient could turn on lights, call the nurse and watch television. I knew that would be easy. I was already controlling my home with a Commodore 64 and X-10, so I thought this would be pretty simple. I wrote a BASIC program for the visual interface and Rex Moses, the Clinical Engineer, designed the electronics.

I wrote the BASIC program and saved it to an EPROM which was plugged into a game cartridge. This allowed the program to be loaded every time the power was turned on. This was faster than loading from a floppy disk and much less expensive than using a hard drive (if a hard drive was even available back then).

Environmental Control Unit [18]

The device was operated by a sip and puff interface that the user would sip to move to the next selection, or puff to select an action. You could change the television station, turn on or off a lamp, turn on or off the computer monitor, signal for a nurse through the nurse call system, answer the telephone or call the operator. The sip and puff switch was connected to the joystick port of the computer.

We packed the computer, the phone and the other electronics into a nice case. We used these for years at the rehab hospital for our patients. Pretty cool!

Unusual Service Calls
The monitor does not work!

One of my first service calls as a civilian at a large medical center in Dallas, was to the ICU. This would have been around 1981 – 1982. The staff had called to report that the bedside monitor was not working. I quickly went to the ICU to troubleshoot the problem. The nurses and doctor in the room asked that I stay outside as they worked on the patient. A few minutes later, they exited the room and said I could check the monitor out. Oh yeah, and they told me the patient had expired! I was still a pretty new biomed tech and I was a little uneasy about checking out the monitor with a dead patient. You think maybe that was the problem?

HDTV and Telemetry

Most medical telemetry from the mid 1960's, to the late 1990's operated as a secondary user of the VHF band – channels 2 – 13. Secondary is the key word here. A secondary operator must yield operations to the primary license holder. You see, the FCC awarded the unused channel for a test site to WFAA, whom rushed the new concept to attract sports arenas and sports bars. That is until one Friday night they decided to test their broadcast.

Medical telemetry uses low power that is typically only strong enough to cover just one nursing floor. The power used by the television broadcasters was high power and capable to reach 50 to 100 miles in all directions. When they decided to 'test' their broadcast, it swamped our medical telemetry and caused a huge problem. No one could see the patients EKG on the screen – it was completely jammed with the HDTV broadcast. Suffice it to say, this was an emergency and our crack biomed team troubleshot the problem looking for the common problem. We worked at it for hours and finally, unbeknownst to us, the TV broadcaster had turned it off. End of test! We 'thought' we had fixed the problem until the next day when they tested it again.

Leadership was very upset that we had NOT fixed the problem the prior day and all hands were on deck to troubleshoot. Luckily, we had a Clinical Engineer, Steve Juett, who had significant RF experience when he used to work at Texas Instruments, and determined what the problem was by was using a spectrum analyzer. He also talked to a local engineering colleague who just happened to know that the television station was broadcasting HDTV on the band we used for medical telemetry (channel 9)! It was a weekend and we tried for a long time to contact the television station to stop testing until we could address the problem. The news room thought it was prank call from a patient at the hospital. Now you know the story behind the new standard that evolved into Wireless Medical Telemetry Service (WMTS)! [19]

And the winner of the Darwin award is ...

Larry Hertzler shared these stories with me.

The first story that comes to mind for me was at my first job in a hospital. One of the older technicians, I think his name was Walt, decided, for fun I guess, to put a sequential compression device around his neck. Of course, it cut off his ability to breathe and he was unable to reach the switch to turn it off. Another tech found and rescued him or he may have won a Darwin award.

During my internship, I learned that a technician at the hospital accidentally came in contact with the main cap in a defib during a corrective service event. He hadn't remembered to bleed down the charge, and it stopped his heart. Luckily, the department was directly across from the ER and they were able to take him there quickly and he made a full recovery. He said he never forgot to test for a charge ever again.

Lastly, for now, and also during my internship in 1980, I came across an old AC defibrillator in a storage room. The energy dial on the front panel showed a picture of a child, a normal sized person, a large person, and the max setting had no picture, but the words written were "Stubborn Patients". Thankfully, I never saw it used.

Are You Wearing Nylons?

Years ago, Paul Stodolka was asked to come up to the burn unit to help on a call about static on a patient monitor. When he arrived, the nurse was mad as hell with him. He found out later the tech had asked the nurse if she was wearing nylons! Of course we know this could be a source of the static electricity but I am sure she did not view it that way!

Percussive Maintenance

This one was sent to me anonymously.

In the late 1970s I worked at a children's hospital. The Pediatric ICU had patient monitors that used a motherboard and a number of daughter boards. We discovered that the daughter boards would sometimes become partially unseated and would cause a particular failure in monitor function. We developed a diagnostic technique that involved a light, open-handed hit at a particular spot on the side of the monitor.

One day I was called into the PICU about a monitor malfunction. After a bit of observation, I began to suspect a board-seating problem. I delivered a noisy but carefully calibrated whack to the side of the monitor and saw the result I expected. However, I had forgotten that our methodology was more appropriate for the shop than the open, multi-bed PICU. Everyone nearby - doctors, nurses, and patients - went silent and looked my way. I turned around slowly and said, "Only members of the clinical engineering staff are allowed to use this diagnostic technique."

Mercury Spill Team

Sphygmomanometers, back in the old days, used to contain mercury. It was not uncommon for someone to drop one and the mercury would go all over the place. We were prepared with an emergency mercury spill cleanup kit. This was great, but I thought you had to dress the part of the emergency response team, so I developed a uniform to go along with the spill kit. Below is a picture of me in my Mercury Emergency Response Team outfit. The light on top was red and would flash and emit a siren sound! Don't you think the Groucho Marx[20] glasses and the Barbie doll, add a nice touch?

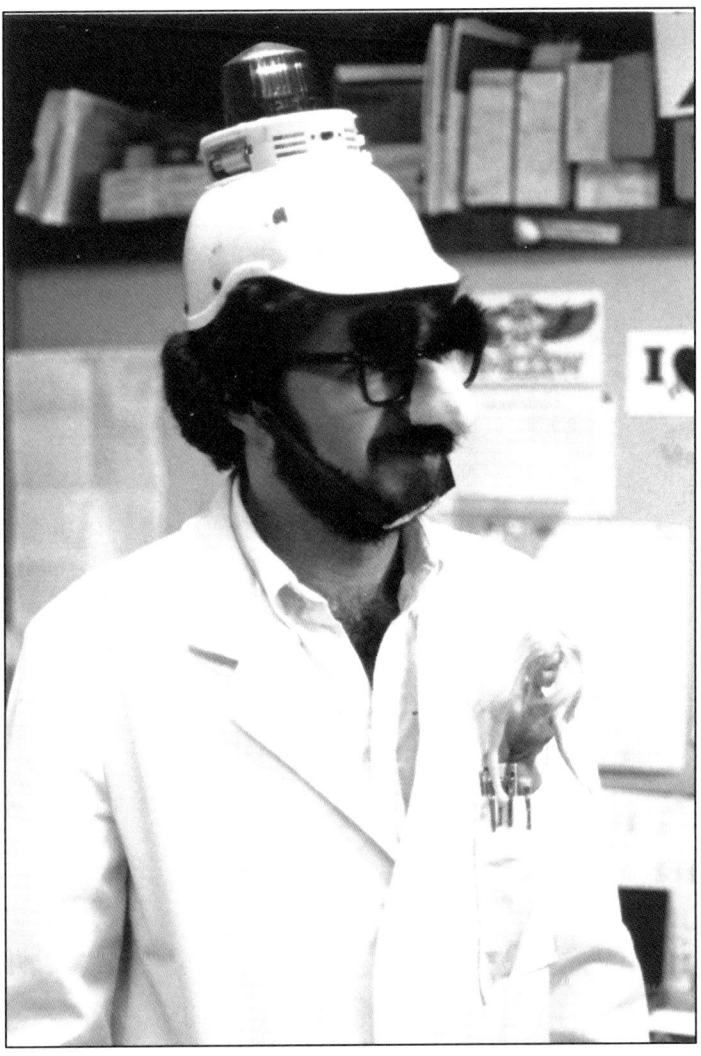

David Braeutigam and the Mercury Emergency Response Team outfit [21]

That god damn Biomed

David Bilder shared this story with me.

Once upon a time, early in my career, one of my first jobs in the biomedical field was at a major medical center in Dallas, TX. For perspective, we are talking about the year 1980. I was hired for the position of being the main Biomed Tech to service the operating rooms of this medical center. At this particular time, open heart surgery was a relatively new thing. And the technology that was evolving throughout all of healthcare, including the technology that would come to make open heart surgery a much safer more routine procedure, was also new and certainly in its infancy.

One particular day, I was called into one of these open heart surgery rooms by one of the nursing staff that I had come to work closely with. She was more than happy to call on me to handle and help out with any of the issues that were related to this new technology and had come to rely on the support of my relatively new Biomed department. To be honest, I don't remember the specific reason I was called into this situation. It could have been a physiological monitor problem, or an anesthesia ventilator problem, or a cell saver problem, or a heart/lung bypass machine. Hell, it could have been that the stereo system in the room wasn't playing

Frank Sinatra as loud as the doctor wanted it to play. All I know is that when I walked into the room, the first thing I heard was, "that god damn Biomed, I'm still working on getting rid of every one of them." Whoa! Not exactly something a 22 year old just starting a job, wanted to hear. After all, I had a $350 rent and a $60 car payment to make. What I do remember is taking care of the issue and getting out of there without getting fired on the spot.

Well, as it turns out, the Doctor didn't follow through with his threat and the Biomed department went on to thrive and to become a very relevant part of the healthcare organization and was an important part of the healthcare delivery team.

I thought about this situation many times in the years to come. Many years later, what I finally came to realize, and what finally hit me, was that this particular Doctor really had nothing against me personally or my department in general. What I do believe is what he was becoming aware of. And that was that his participation in these super technological advancements were possibly not what he wanted to see and he was not going to put the effort in to learning. This is really not a criticism of the Doctor, but more of what was happening at the time. The Doctor was reaching the end of his career and I was just starting mine.

For me, to live and breathe the medical technology advancement, to live, breath, touch, feel, smell and work on what has evolved over the past 30+ years, I still find amazing.

Ouch!

Here is a memorable call from Doug Dreps when he was a biomed tech back

Another memorable story back in the early 90's, we received a call that a Nellcor N-200 probe was hurting a patient's finger. Responding to the call it turned out no probe was hooked up, but the patient module that had teeth on it to grab hold of a blanket was directly on the patient's finger. Not wanting to upset anyone I pulled the nurse outside of the room and explained what went wrong. This is what we have done on the spot over the years, direct caregiver training on the equipment they use. Thankfully, we were called right away, so no harm was done.

Shocking!

Just after I was promoted to a Supervisor position, a call came in from the Physical Therapy department. The caller stated that the LifePak 5 defibrillator was not working. The tech that took care of that area, Don Hildebrand, was on another call. No problem I thought, I will just take the call myself. I went to Physical Therapy and located the LifePak 5 defibrillator mounted on the wall. I took the paddles out of the holders, placed one in each hand and push the energize button. I immediately started to receive a shock. Somehow I was able to throw the paddles out of my hands. The therapist quickly came to my aid and applied hot towels to my hands. We found an entry and exit wound the size of a pencil lead in each index finger of my hands. I called down to the shop and had a tech come take the defibrillator out of service. It happened to be Don Hildebrand. I told him which defibrillator it was and I still remember him saying "Oh yes, I can smell something burning." My reply was "That would be me!" After calming down, I went to the Emergency Room to get checked out. They did an x-ray and a twelve lead EKG. No problems were found. So they bandaged each index finger and sent me home. Since I had been in the Emergency Room for several hours I didn't arrive home until probably 6 or 7 that night. I hadn't called my wife (pre cell phone days) so you can imagine her surprise when I arrived home with each index finger wrapped with very large bandages. She thought I had just lost both fingers!

We had the defibrillator checked out by our staff and could find nothing wrong with the unit. We sent it off to the manufacturer and they could not find anything wrong with the unit. My best guess was there must have been residual conductive gel left on the paddles that created a conductive path to my fingers. This was before the use of defibrillator pads so you had to use conductive jell to have good contact with the patient. Needless to say, I carefully check defibrillators before testing them out in the future and made sure there was no residual conductive jell on the paddles.

Was MacGyver[22] a biomed?

Here is one sent to me by Richard Eliason.

One of the myriad of things I enjoy about being an HTM professional is that every day is different. Each day brings new and often unexpected "opportunities". But on a day, many years ago early in my career, I wouldn't have ever guessed I would receive a specific request from a veteran CRNA to "wire a cooler". Yes, the same vessels produced by "Coleman", "Igloo", "Yeti", and the like often used to chill your beverage of choice while enjoying the great outdoors or while pre-game tailgating on a weekend.

After realizing the request was NOT a joke, I asked the CRNA to elaborate on his request. He looked at ME like I was crazy for not immediately recognizing the serious nature of his request. He sternly pointed to the drain hole in the cooler, explaining to me that this is where I should run the power cord. Really? Yes, he insisted, just like the one in the Anesthesia workroom in the OR. With his last statement I politely asked him to show me the referenced cooler located in the workroom.

Upon arriving to the workroom, the CRNA pointed to a remote corner where there was a Styrofoam cooler (Undoubtedly the new cooler would be a modern upgrade in his eyes) filled with IV Bags of Mannitol[23]. I could clearly see a hole had been punched through the side of the cooler where a power cord exited and was plugged into the adjacent wall outlet. Intrigued by the situation, I begin to carefully remove the IV Bags from the cooler until at the bottom of the Styrofoam cooler I revealed, to my astonishment, a 5"X10"X3" wire mesh cage to which a 50W ceramic resister was soldered. Attached to the respective ends of the resistor were the "hot" and "neutral" wires of the power cord. The ground wire from the power cord was connected to the wire mesh cage. An ingenious make shift "warmer" to keep the Mannitol from crystallizing in which MacGyver would be proud to be sure! Ingenious, YES, but a potential death by electrocution situation waiting to happen should an IV bag burst or unwittingly someone reach in and grab the exposed 120 Volts!

That day I temporarily lost favor with the CRNA when I refused to "upgrade" his cooler AND informed my Director and our Risk Manager of the current situation. After the fallout, the Styrofoam cooler was removed from the workroom for the safety of all and a proper UL approved warmer (that passed electrical safety inspection) was purchased specific for its intended use.

I never found out (and no one would admit to) what "creative" mind constructed the "cooler of death", but to this day every time I see a cooler I am reminded of this story and how the most mundane of days can turn into a biomed "adventure" AND what an essential role BMETs play in keeping hospital staff and patients safe.

I hope you know CPR!

When defibrillators first came out with non-invasive pacing the sales team of all the major defibrillator manufactures' would come to the biomed shop to demonstrate this new feature. One particular sales person from Zoll was going to demonstrate this new option to Richard Swim and me. He opened his shirt, applied the pacing pads and as he was getting ready to pace himself he asked if we know CPR! Both Richard and I said we are not going to touch you! This is crazy! The sales guy just stood there pacing himself all the while talking to us about the new feature!

I shared this story with some Zoll representatives many years later and they stated that all the sales reps were required to do this. Talk about a tough sale!

Hey Dave, Got my Blade?

One day after lunch we were coming back to the shop to play darts. I think it was Forrest Parker, Richard Swim and I. (It was a simpler time back in the 1990's – we weren't as concerned with productivity as we are now.) The director of the ED happened to see me and yelled out, "Hey Dave, got my blade!" He was referring to a laryngoscope blade I had picked up earlier but it sounded like he wanted to know if I had his knife! We all got a good laugh from that and then proceeded to play darts.

Priceless

Another story sent to me by Doug Dreps.

Another story from the 1980's, was the old Anesthesia HP units in the OR that had enough knobs on it to scare most clinicians, yet alone some biomedical technicians. One of the Anesthesiologists was also an EE (Electrical Engineer) and he understood the equipment very well. He understood electronics and was very good with medical equipment in the OR and ICU. He took a much needed vacation one summer. I received a panic call from other anesthesiologist's interns that they were moving some of the knobs on the monitor and could no longer see what was needed. They pleaded for me to help them as our vacationing Anesthesiologist was due back the next week. I came over and made the necessary adjustments and told them I would not say anything. The joy in their faces was priceless.

Bunny and the OB-Gyn exam chair

I once got a service call to check out an Ob-Gyn exam chair at a clinic in the Hurst-Euless-Bedford area of the Dallas-Ft. Worth metroplex. The nurse's name was Bunny. Bunny told me that sometimes when a patient is in the exam chair and the doctor repositions the chair, it freezes up and the chair will not move. This is both awkward for the patient and embarrassing for the doctor. Bunny and I both operated the exam chair numerous times but could never get the exam chair to fail. I closed the service ticket and went back to the office. A week or two later, Bunny called back and said it was failing again and the doctor was furious that I had not fixed it the first time. I went back to the clinic and was trying to decide how to best simulate the problem. So I just got into the exam chair, placed my boots into the stirrups (I do work in Texas!) and Bunny operated the controls until we finally got it to fail. I was finally able to determine that the large bundle of wires that ran the length of the exam table, had an intermittent break. I ordered the new cable and quickly got the exam chair back into operation. As a bonus, Bunny and the other nurses got a good laugh by watching me sit in the chair like I was getting an Ob-Gyn exam!

The belly dancer and the monkey

A week or two after I started my job at a hospital in Dallas, David Bilder was celebrating his birthday. Our boss, Jeff Butler, thought it would be a great idea to celebrate. He hired a belly dancer and an organ grinder with a monkey for the celebration! Here I was in the middle of a crowd of people, some I did not know yet, watching the belly dancer and the organ grinder and his monkey. Finally a lady from the lab turned to me and said "I just came here to pick up some equipment." I am sure she was as surprised as I was at the birthday celebration!

Flasher

I was on call one summer and was called into the ICU after for a balloon pump issue. I was wearing a polo shirt and shorts. It was quicker to go in to the hospital instead of going home to change. Besides, we wore white lab coats back then so I would just put one on.

I came into the hospital, went to the biomed shop and put on my white lab coat. I then went directly to the ICU to find the nurse that called me in. She looked at me and asked what I had on. I said "Shorts." Then I realized the lab coat was hiding my shorts and it didn't look like I was wearing anything on underneath. Boy was I embarrassed!

Wussy [24]

I got a call to the Physical Therapy department to work on an electrical stimulator. It was a hand held portable unit with 60 volt output. The therapist said it did not work. I connected myself to the unit, slowly cranked up the output, wiggled the wires and all of a sudden I got the #@%!$@# knocked out of me! I thought "Must be an intermittent wire." I repaired the wire and checked the output on my multi-meter this time and it worked great. I took it back to the therapist and she asked if it was fixed. I said yes. She asked me if I tested it and I said not on myself because I had gotten shocked earlier. She looked at me and called me a "Wussy!"

Jack of all trades

Doug Dreps remembers when.
I remember back in the 1980's a nurse called and told me the jack was broken on an Ohio infant warmer. I went up to the unit and saw about 2 feet of nurse tape on the probe holding in the jack. I had worked on these units for years and never saw a jack go bad. After taking all the tape off, I saw it was an Airshields infant warmer probe, which is a smaller diameter. The good news was it wasn't broken. The nurse took the news well.

Captured in Tehran

Doug Stephens is well known in the biomed community. He is known both for his company, Stephens International Recruiting, Inc., and for his commitment to biomed certification. Did you know he was also captured in Tehran during the late 1970's? An article published in TechNation Magazine in September 2017[25] tells the whole story. I will summarize it here.

Doug Stephens had just been promoted to Chief Warrant Officer 3 when he was transferred from Vietnam to Tehran, Iran in 1978. His job was to provide biomedical, plant operations and communications support to the U.S. Embassy in Tehran. Tehran was already an intense environment, especially if you were American. It was not uncommon to see anti-American graffiti written throughout the city. They were told to be careful but Doug, his wife Cindy and their daughter felt they were safe. Doug had a rifle pointed at him one time on the way to work. Another time a Molotov cocktail was thrown through their window. By late 1978 the situation was worsening. By the end of 1978 American families were being evacuated for their safety. Cindy had to stay since she worked at the hospital. Finally in early 1979 Cindy and her daughter, Dianne, were evacuated to safety but Doug stayed back.

Then the unimaginable happened. A revolution in Iran started on the 14th of February 1979. On the 15th the Islamic Revolutionaries entered the compound were Doug worked and took everyone prisoner by gun point. They were interrogated, searched and lined up against a concrete wall. Needless to say they were concerned for their safety. The revolutionaries painted "Down with Carter" on the wall. They were told to stand there for four hours.

No one was allowed in or out of the compound. All communication was cut off. Doug, being a resourceful BMET, quietly hooked up a phone in his office so he could stay in touch with his wife. By June 1979, ten of the captives were released after the U.S. Embassy negotiated with the revolutionaries. Doug stayed back to collect the medical files of the American staff and send them to Germany. He was released on July 18, 1979.

The alarm would not stop

Doug Dreps remembers this one.

We often receive notes on equipment that say "broken" or "fix me". One time we had a note on an infusion pump, saying the alarm would not stop. Someone had taken surgical scissors and stabbed through the cover into the speaker to get it to stop alarming. Luckily, they did not get shocked. A minor repair turned into a case and speaker replacement.

Sticky Notes and Nurse's Tape

Equipment typically comes to biomed with a sticky note or nurse's tape and the 'failure' written on it. It would be great if we would get something more descriptive than just 'broken'. Occasionally we will get a funny or memorable note written about the equipment failure.

I've heard of several shops that even had a wall with the funniest sticky notes displayed.

 Here are a few interesting notes found on equipment sent to me.

"Ticks like a bomb!" Even I know what to troubleshoot! [26]

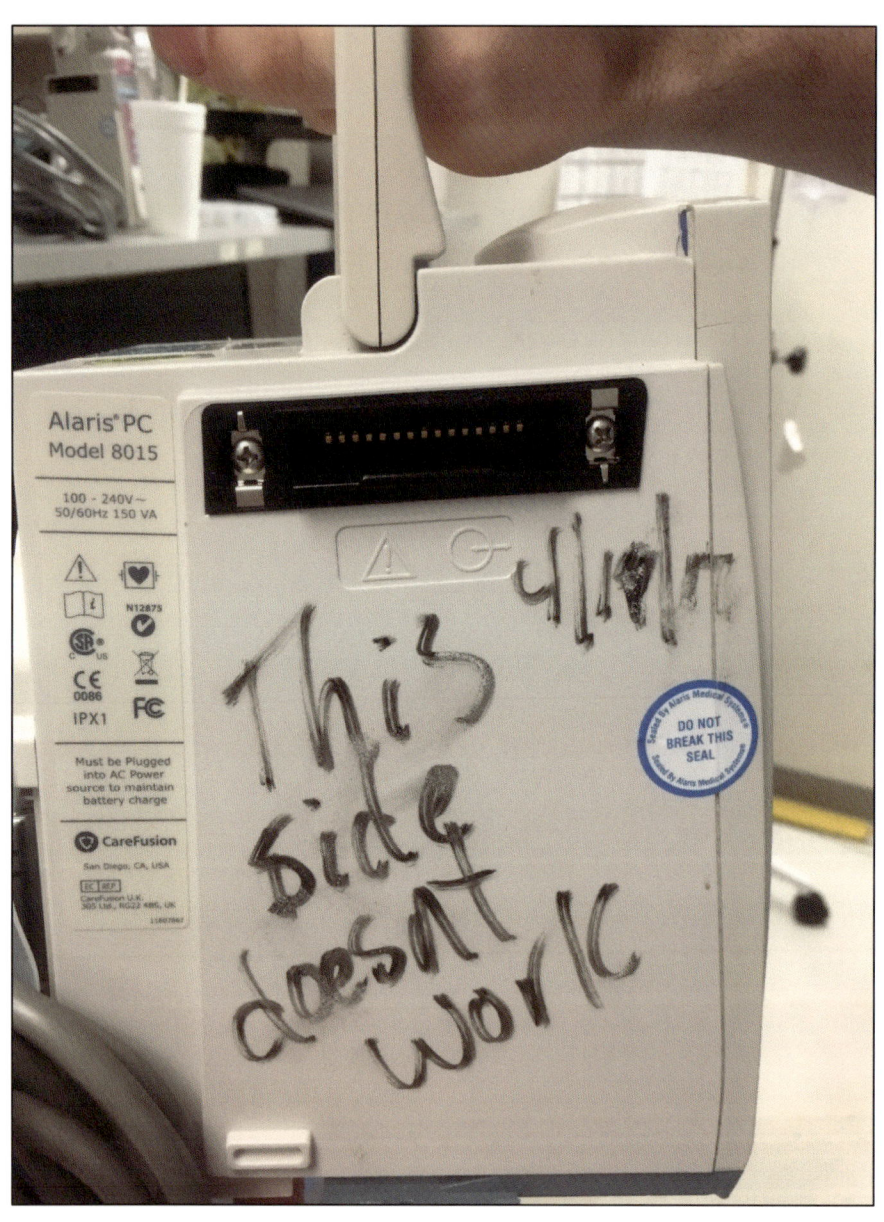

No sticky note needed – just use a sharpie! [27]

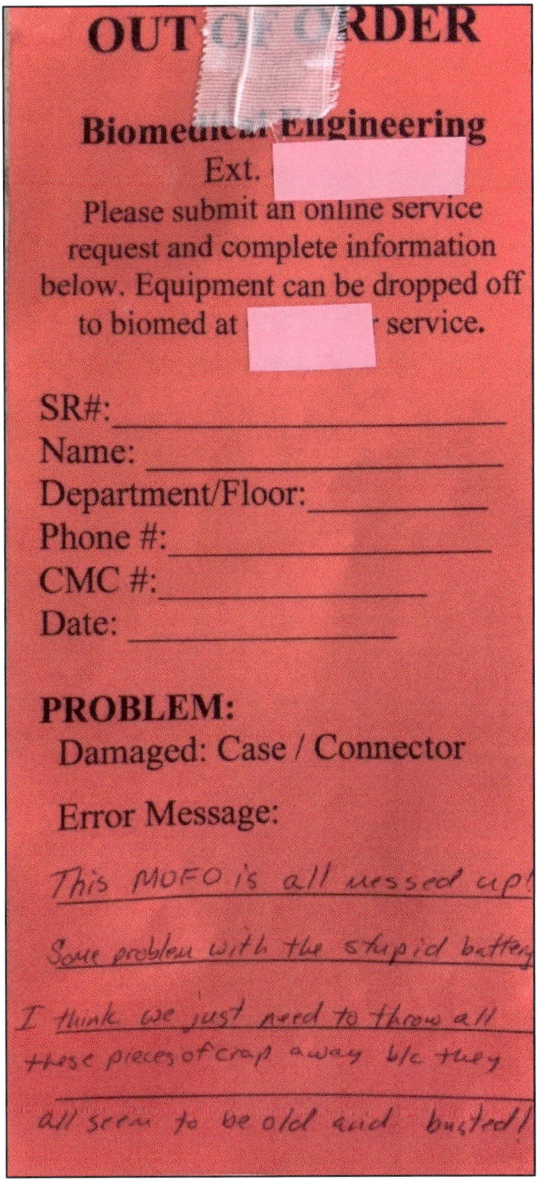

OUT OF ORDER

Biomedical Engineering
Ext. ▓▓▓
Please submit an online service request and complete information below. Equipment can be dropped off to biomed at ▓▓▓ service.

SR#: _____
Name: _____
Department/Floor: _____
Phone #: _____
CMC #: _____
Date: _____

PROBLEM:
Damaged: Case / Connector

Error Message:

This MOFO is all messed up! Some problem with the stupid battery. I think we just need to throw all those pieces of crap away b/c they all seem to be old and busted!

I don't think they like this equipment! [28]

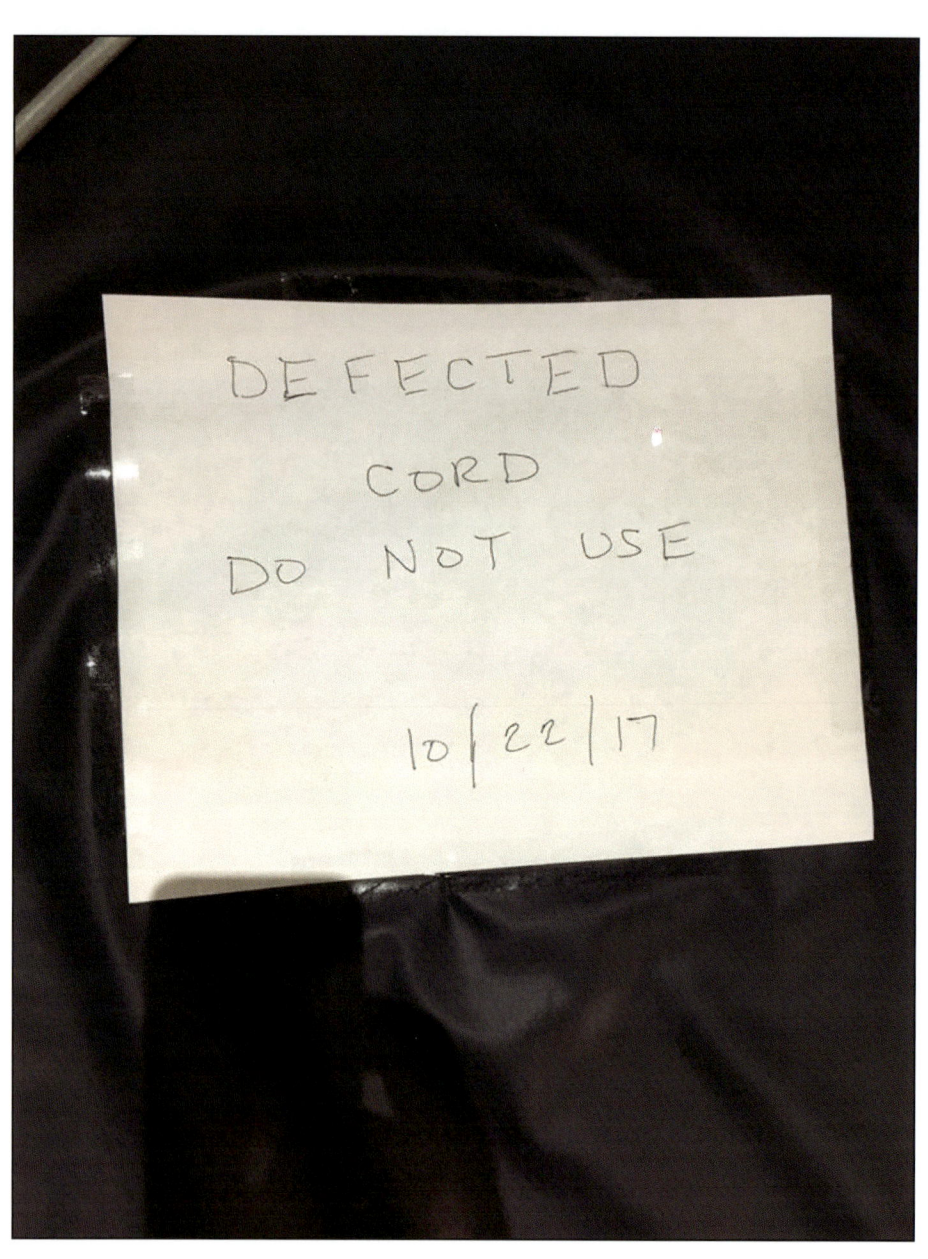

This cord did not have a visa! [29]

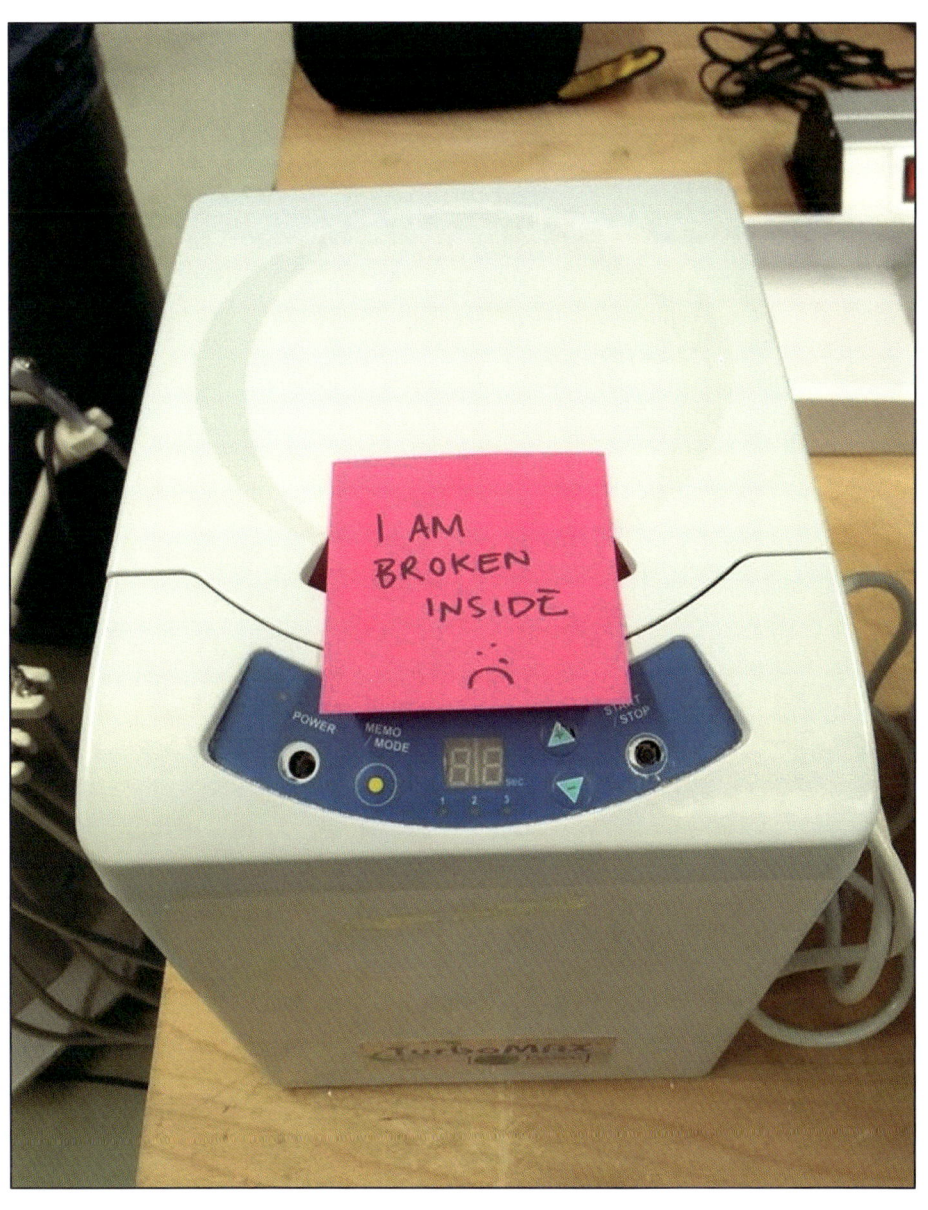

A cry for help [30]

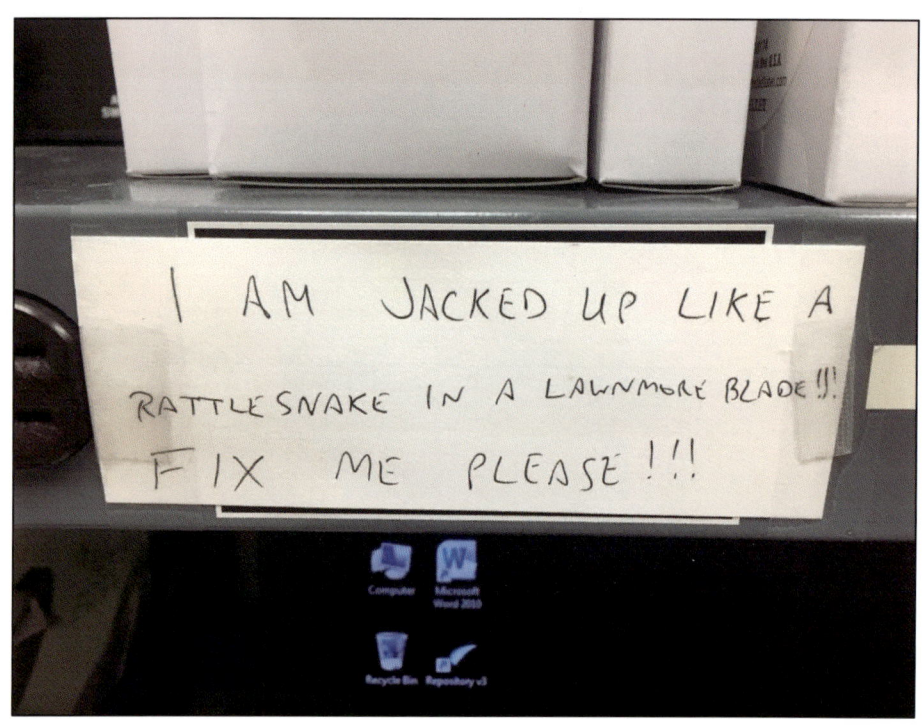

A rattlesnake in a lawnmower blade? [31]

I think the problem is the NIBP hose is grounded! [32]

The chaplain might be better for this than a biomed! [33]

"Beeps Occultion"

Bacon tape for a Centrifuge [34]

For the love of God please fix this! [35]

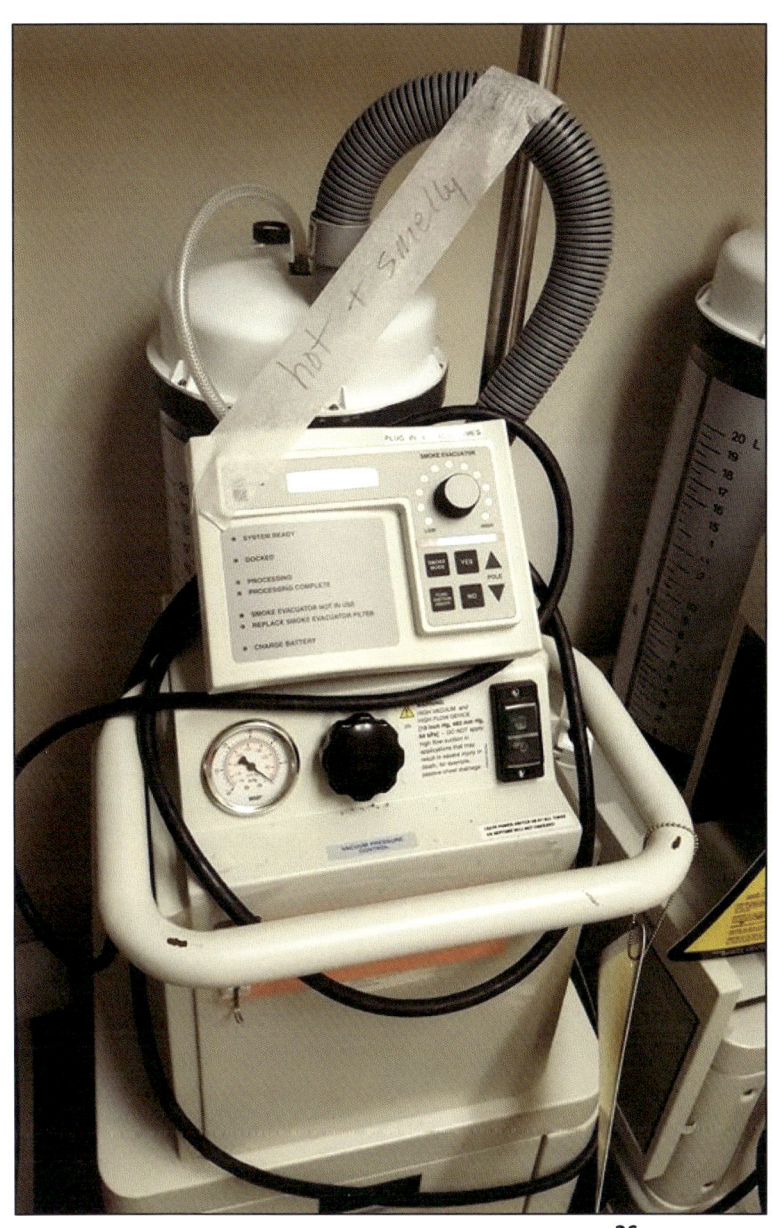

Aren't we all sometimes? [36]

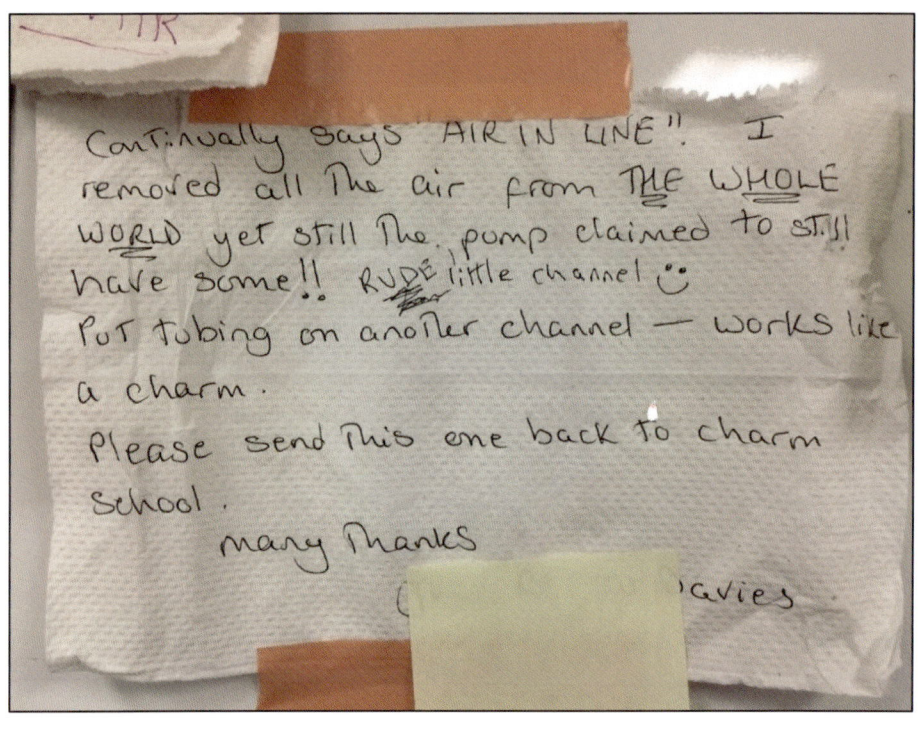

This note actually gives all the detail you need! [37]

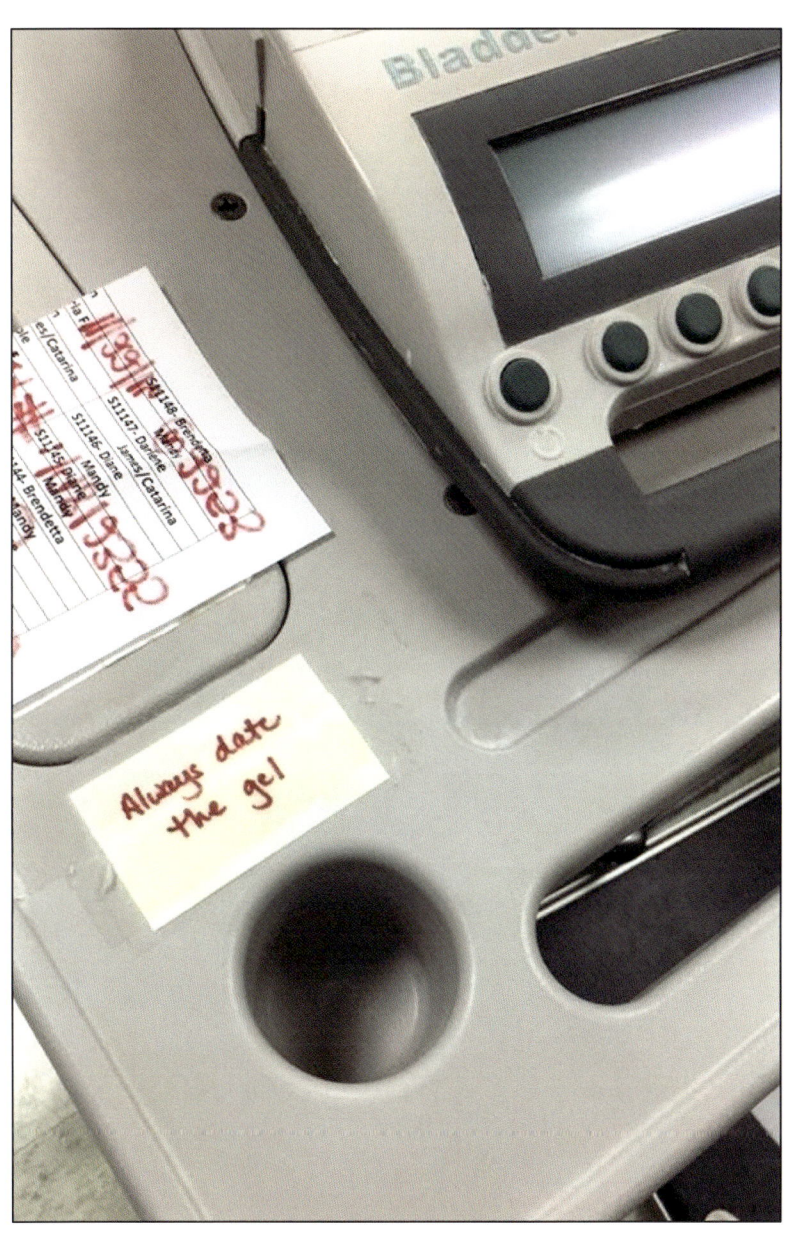

For all you singles out there … [38]

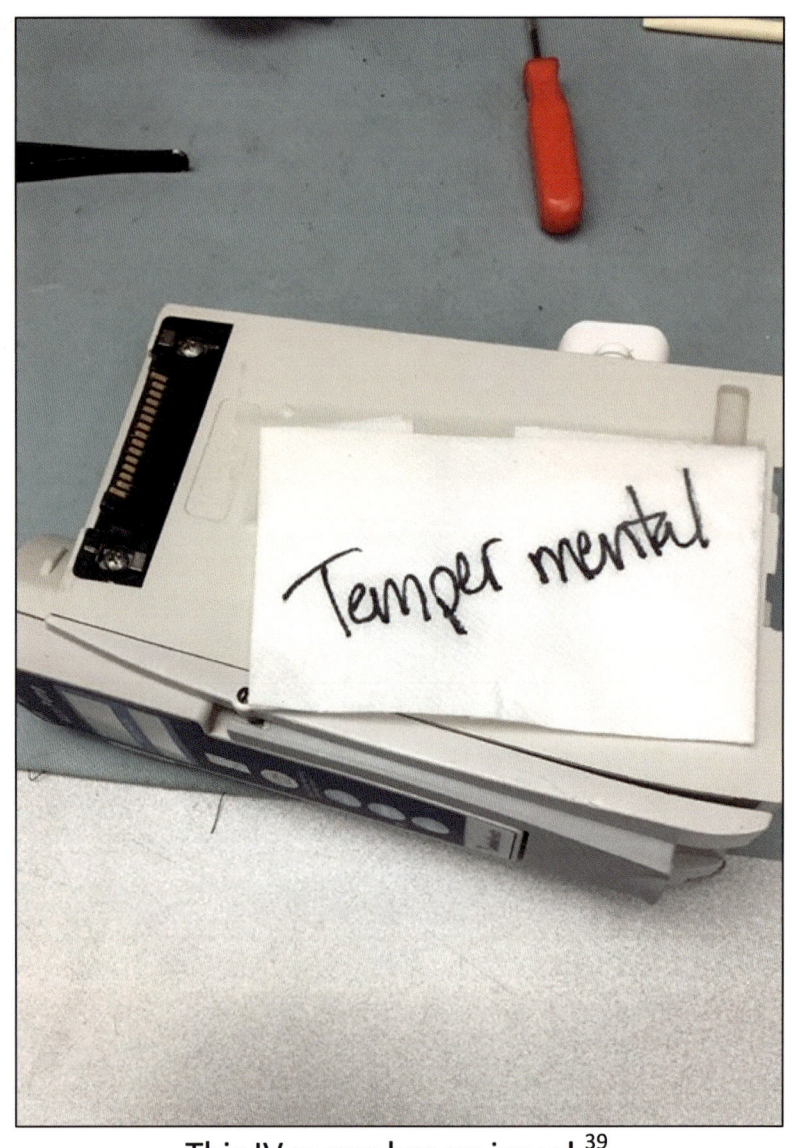

This IV pump has an issue! [39]

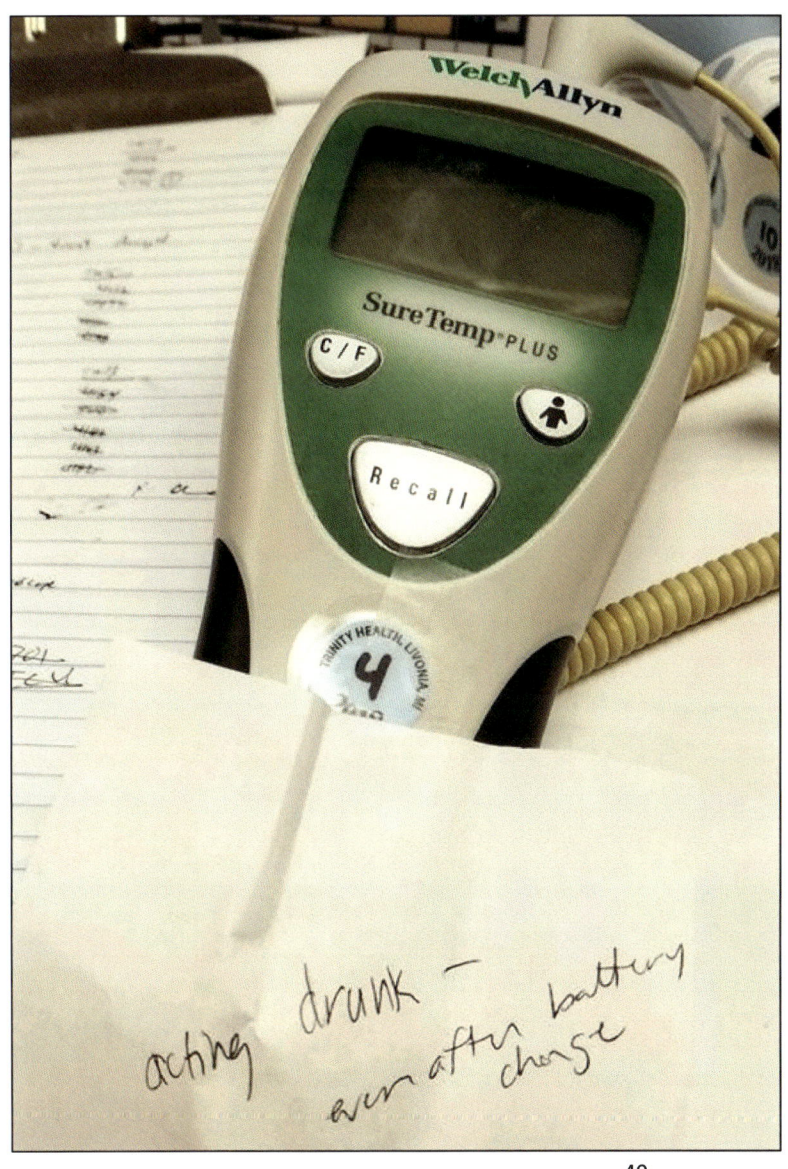

Are you sure it was only acting? [40]

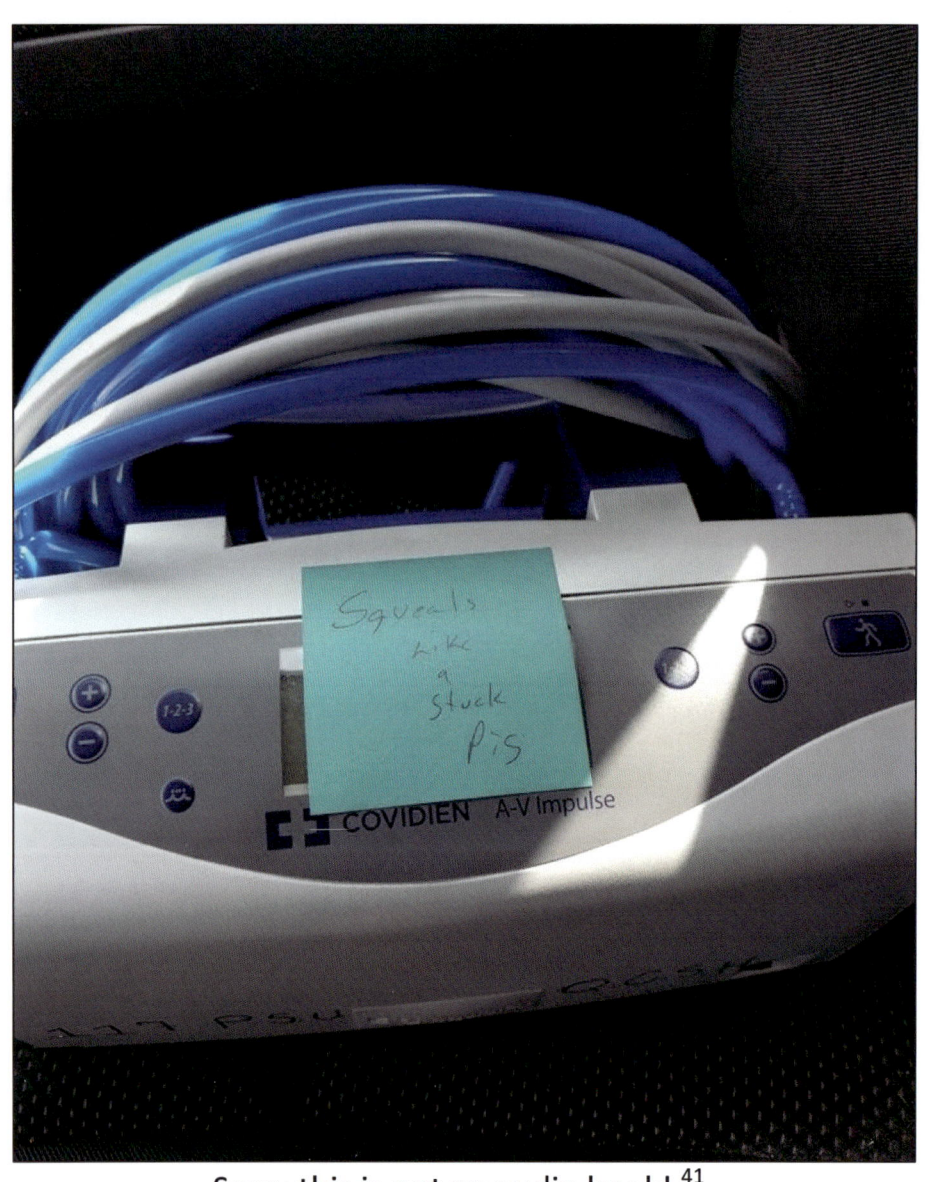

Sorry this is not an audio book! [41]

"Squeals like a stuck pig?

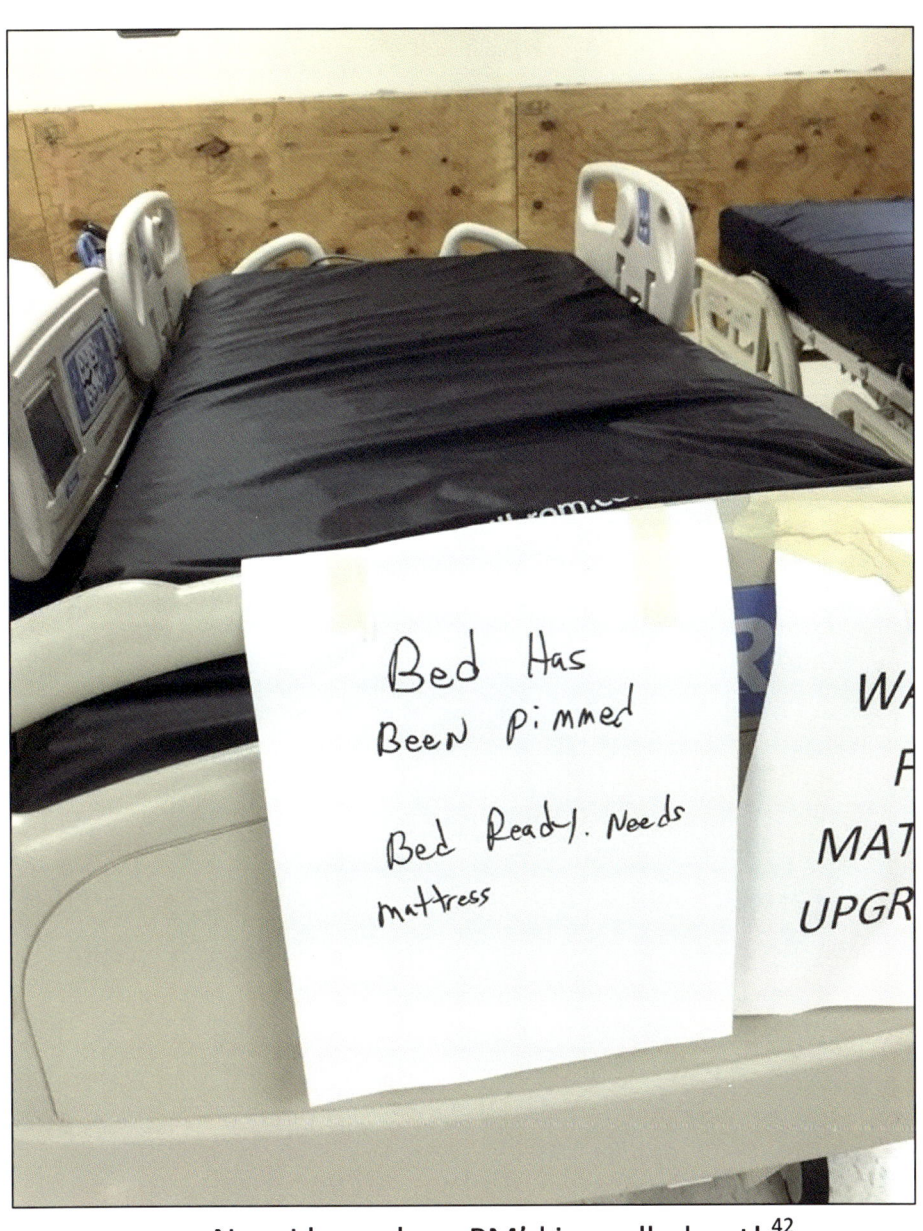

Now I know how PM'd is spelled out! [42]

Are they just mad or is there a problem? [43]

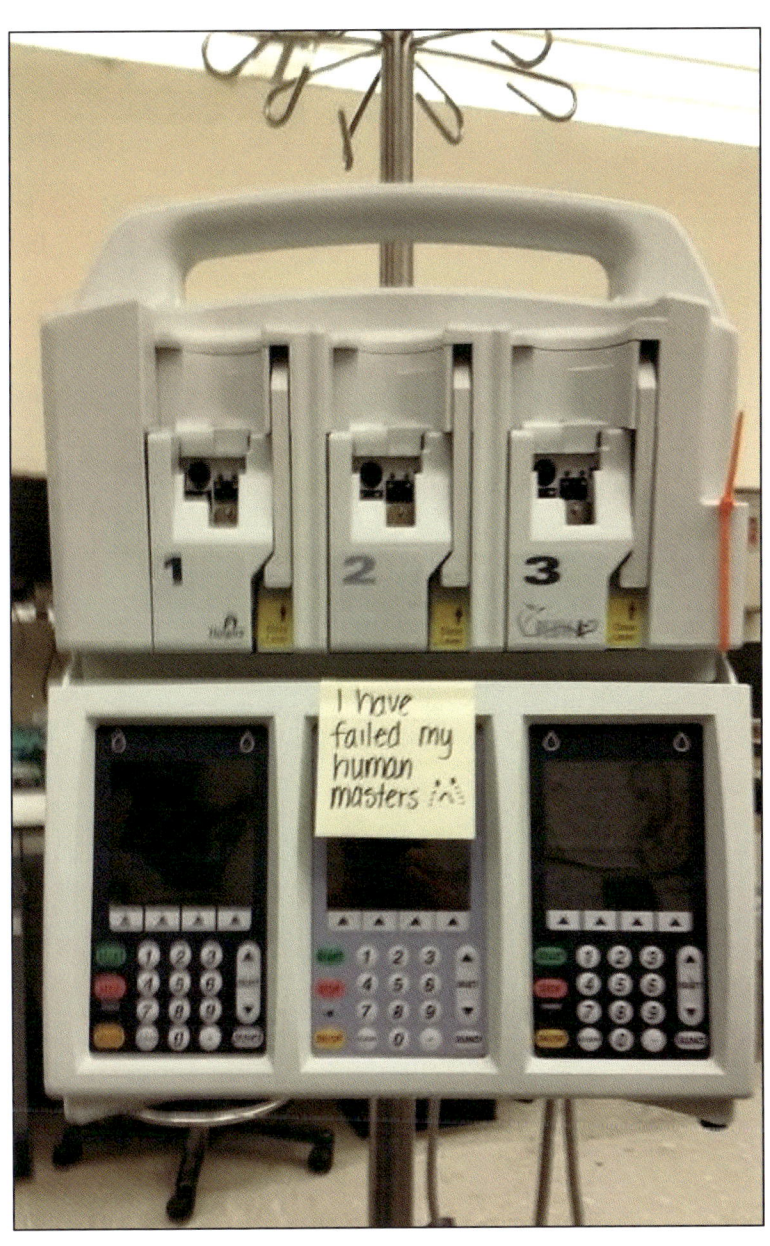

A sense of humor! [44]

Did it go back to where it was manufactured? [45]

I'm not happy and neither is my brother! [46]

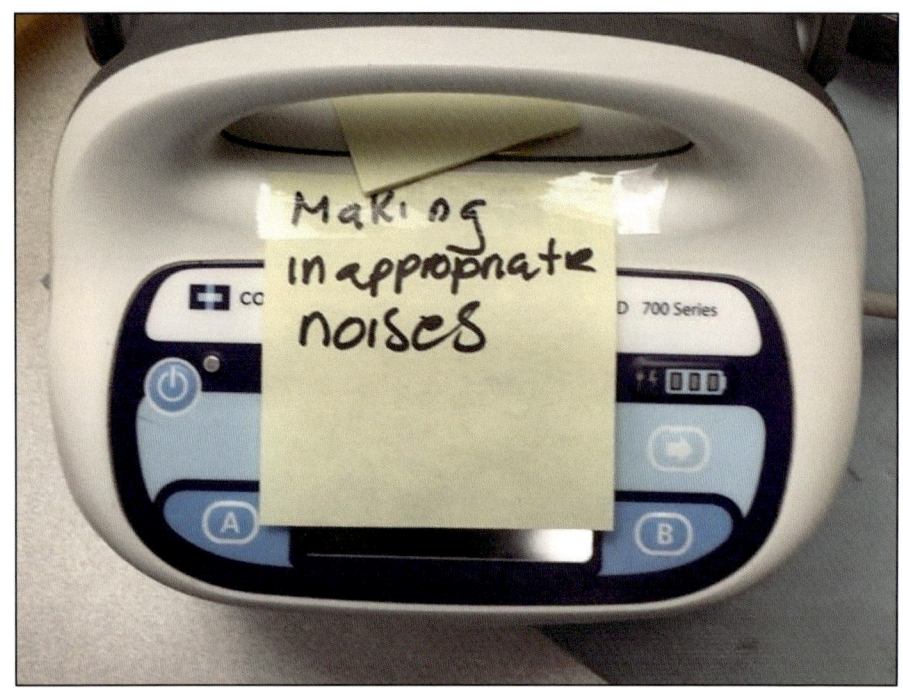

Probably should call HR and not biomed. [47]

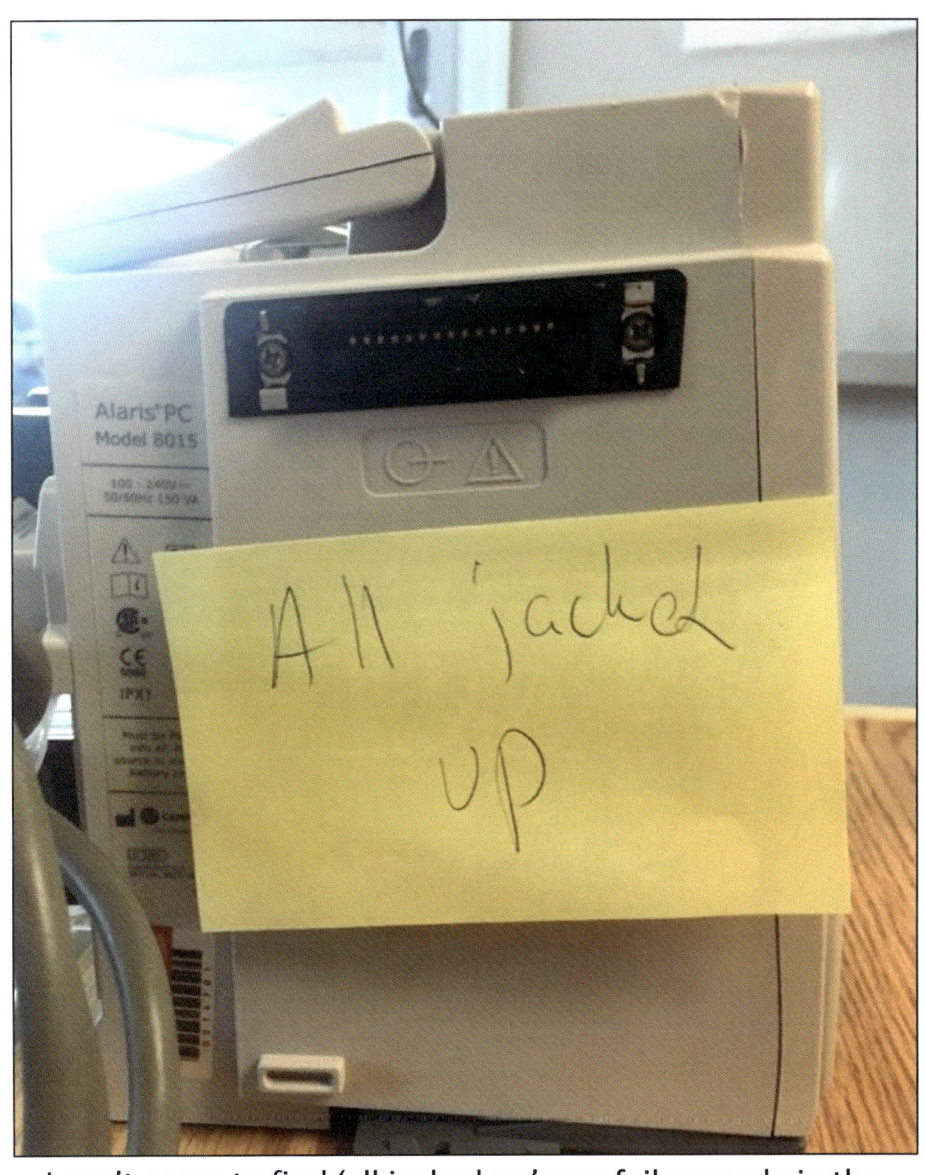

I can't seem to find 'all jacked up' as a failure code in the service manual. [48]

Looks like the night shift was bored [49]

Biomeds are a Resourceful Bunch
Camping in the Army

While stationed at Fort Bragg, North Carolina, for the 32 MedSOM (Medical Supply Optical and Maintenance), our job was to support the 82nd Airborne and other troops if they were ever deployed to go to war. Part of our job was to practice these events so we would be prepared. My first practice event was a two week affair in the panhandle of Florida, just outside of Eglin Air Force Base. Lucky for me we had seasoned biomeds that had gone on several of these before so they knew how to prepare. Since biomeds in general, are very resourceful, we designed our own private shower using sheets of plywood (painted green of course), procured a shower head and a large 50 gallon barrel with a submersible heater. We now had our own private shower with heated water. We didn't have to worry about going to the community shower like all the other guys.

 We were also set up to provide blood storage in the field. This meant we had a blood refrigerator to take with us. This came in very handy to store our soft drinks and other food that needed to be kept cold. Needless to say our 'camping' exercise was pretty easy compared to everyone else. I wonder what we did with the blood?

Honeysucker

During another field exercise while I was stationed at Fort Bragg, was to manage the biomed techs under me. I had a Private Vernetti who reported to me. Vernetti was a large, physically fit guy from upper Wisconsin. This guy could lift anything. One assignment in the field was to suck the septic waste from our porta potties. I assigned Vernetti and another private to take on this glorious task while I managed them. It helps to be a sergeant in the Army! Vernetti just couldn't do it. He gagged at the thought of human wast being sucked up through that hose. So I took the hose myself and 'sucked' the waste into the honeysucker.[50] You could 'feel' and hear the human waste being sucked into the truck. It had to be done and I did it for Vernetti! Yuck!!

A Beacon of Light

In the mid 1980's, we took care of computers for our hospital. We were working with Information Services to install a new computer system. Like most projects, this was delayed, and then delayed, and then delayed for various reasons. Finally, the night came, after almost a year, we were going live. Information Services printed commemorative t-shirts to celebrate the success. It was a light tower showing a beacon of light projecting from it with the go-live date. It was a very impressive t-shirt. Biomed decided to create their unique t-shirts also. Our t-shirts had BCON go-live with each of the go-live dates marked out on the t-shirt. Leadership in Information Services did NOT like our t-shirts but they sure were popular with everyone else, including the IT staff!

The Plague

While stationed at Walter Reed Army Medical Center in the U.S. Army in the late 1970's, we worked on all kinds of medical equipment and in a multitude of environments. Since Walter Reed was a military hospital and also performed research, we worked around anything they were researching, including the Plague. Yes, the Plague – like Black Plague, Bubonic Plague, etc! For the biomed techs at the hospital we had to be immunized against the plague. Hence my shot record from the military showing I was properly immunized! I wonder if this is a violation of HIPAA?[51]

PERSONAL HEALTH HISTORY

The information which follows is a record of other immunizations which the traveler has obtained as an additional health protection for international travel. These immunizations are NOT usually required for entrance by any country. Space is also provided for a personal health record in case of illness or accident while traveling abroad.

OTHER IMMUNIZATIONS (Typhus, Typhoid, Plague, Poliomyelitis, Tetanus, etc.)

Date	Vaccine	Dose	Physician's Signature
4 Apr 77	tyh	.5 cc	GARETH E. BROOKHART WO1, PA, 507-36-3450
4 Nov 77	tet	.5	GARETH E. BROOKHART WO1, PA, 507-36-3450
MAR 9 1979	Plague	.5	GARETH E. BROOKHART WO1, PA, 507-36-3450
NOV 28 1979	Flu	.25	THOMAS L GREGG, PA-C Hollingsed, PA-C
AUG 21 1980	Typh	0.5	Michael J. Hollingsed, PA WO-1 463-84-2017 MC 138-88-3459
AUG 21 1981	Tet	0.1	WO-1 MC 436-88-3459
NOV 5 1981	Flu	0-sc	KENNETH D. GOOGE CW2, PA, 261-84-7640

David Braeutigam U.S. Army shot record [52]

Fitzsimons Stories
Taco Bell and the Marines

I remember when I was stationed at Fitzsimons Army Medical Center[53] in Aurora, Colorado going to the short course, probably around 1977. Several off us went down to a Taco Bell on Colfax Avenue. I was standing in line with everyone else getting ready to order, when a VERY large man approached me and asked me if I was a Marine (remember, this is the late 70's and everyone had LOTS of hair back then except those in the military). I stated "No, I am in the Army." He said, "That's too bad, I'm looking to kick some Marine's Ass!" I was never so proud to be a member of the U.S. Army!

LifePak 33

Another story at Fitzsimons I remember, was during a practical test. The instructors would place a 'bug' in a piece of medical equipment and then see how long it would take you to find and identify the problem. I believe you had 15 minutes to correctly identify the problem. The instructors would walk around the class during the test to see if you needed help and then would pass you upon successful completion. This day I had a Physio-Control LifePak 33 defibrillator. (Trivia – the LifePak 33 was the first portable DC defibrillator. It was named the LifePak 33 because of its target weight of 33 pounds. Two units were purchased to place on Air Force One and Air Force Two).[54] I seem to remember there was a problem with the CRT display. I turned the power off (I thought!), reached inside the unit to remove a board to plug into an extender card, and promptly received the high voltage current that was feeding the CRT. Needless to say, I let out a big scream upon receiving the shock. My instructor came by and kindly told me to "just relax." No @$@#@#!

Club Fitz

This story is shared by Thomas Putt.

 Wintertime at Club Fitz shop in Colorado in the late 80's. Leadership decides to put lock boxes on the building thermostats cuz the belief is the spoiled BMETs are using too much taxpayer steam heat. One of the occasional sub zero blizzards has blown in and everybody is freezing in the shop, which is set for a brisk 60 degrees, but is actually heating to the low 50s. Military and civilian personnel have shed their tissue thin medical whites for winter weight BDUs or heavy pants and sweaters. Orders are to leave the thermostats on their current settings and "work harder and you'll all get warm."

Interesting theory, but the spoiled BMETs have a Plan B. Several cans of canned air are enlisted to help remedy the oppressive and cold working conditions. With a little coaxing, the canned air cans invert themselves and discharge their cold, liquid propellant through their straws between the plastic slots on the thermostat covers, thus drenching the thermostat sensors to a much lower temperature than the already cold ambient air.

Proof of a successful operation comes by way of the creaking and cracking sounds of the steam radiators as the precious taxpayer steam is released.

All is well and the shop is warming nicely. Soon, however, shop personnel notice perspiration forming on their foreheads. Plan B has betrayed them. Quick check of the shop temperature is low 90s and radiators still cranking. About the time temps are mid 90s, CW3 Sine comes out of the office yelling "Why is it so #@%@#* hot in here?" Chief realizes immediately everybody's involved by the in unison turning of backs to him, while stifling laughter.

Chief checks the first thermostat box, finds it secure and manual setting intact to 60 degrees. Chief says "Great, I don't have keys to the box. Whatever you did, fix it. Now." Chief grabs his jacket and cover and leaves.

Temp in shop is passing 100 degrees and everybody is scrambling. Windows and doors are wide open, snow is coming in pretty heavy, but melting immediately upon contact, so we've got puddles forming everywhere.

Heat Guns are being run under the thermostat boxes as rapidly as possible. Interestingly enough, the plastic boxes are acting as a heat sink to save the precious thermostat sensors. Not enough heat guns for the number of thermostats, so cig lighters are enlisted to correct the temperature discrepancy. Soon, the smell of burning plastic is becoming noticeable.

About 45 minutes pass before the steam radiators are not creaking anymore and beginning to cool. Thermostat covers are a bit misshapen, but still serviceable. Temp is down significantly and puddles are mopped up.

Chief returns with keys and new shop orders, "No #@&@# with the thermostats. Period. In fact, don't even look at them."

Chief takes the thermostat covers off and gives 'em a quick eyeball, then sets the temp to about 72 and locks the boxes. Hey, operation successful! That wasn't so hard after all! Sometimes the process is more important than the result.

Sterilizer bugs

If you finished all the 'bugs' in the equipment the instructors at Fitzsimons would play games with you. It was my turn to troubleshoot a large sterilizer and it was their turn to play a game with me. After turning on the sterilizer and determining the problem - I proceeded to open the door of the large sterilizer. Out of the sterilizer came all the water that had filled the chamber. The water flowed onto the floor and the instructor quickly pointed to a mop and told me to clean the floor. Lesson learned!

No Smoking Please

During this time (circa 1978) the military was just starting to accept separate smoking and non-smoking areas. The break rooms at the school had just been changed to smoking and non-smoking areas. I was a very big anti-smoking advocate. I was also aware of the new policy of the Army and realized they did NOT offer non-smoking areas in the cafeteria. I followed my chain of command and requested one be set up in the cafeteria. Eventually this happened and I was excited to sit in the new non-smoking area once it became available. Once, I sat down, and soon realized they also seated all the patients that were shot up with Thorazine[55] in this area! Yikes!

Let's be Cool

During physical education at Fitzsimons, all the smokers would roll up their t-shirt sleeve with the packet of cigarettes inside. I did not smoke but did not want to be left out of this 'cool' act. I went to the store and bought a packet of candy cigarettes so I could role them up in my sleeve along with almost everyone else. I was now 'cool' like everyone else! Plus I didn't have nasty smoker's breath!

Wax On - Wax Off

Barrack's life in the Army at Fitzsimons was pretty interesting back then. One of our duties was to wax the floors to a very high sheen. You accomplished this by using a motorized buffer and wax. I don't recall if we ever receive formal training on how to use the motorized buffer but surely it couldn't be that difficult. I remember the first couple of times I operated the buffer, it was extremely difficult to keep the buffer from not going side to side so fast that you would crash into the wall. Either someone finally showed me how to properly operate, it or I finally figured out that you had to just gradually tilt the buffer ever so gently for it to respond correctly. I still remember how hard it was the first couple of times and how hard I had to work to get it to wax the floors. I was exhausted until I learned the proper way to carefully use the buffer.

Boston Baked Beans

Anyone who attended Fitzsimons remembers getting their mail after school and going through the swinging doors to the outside. I always wanted to prank someone using those doors. Finally, I had the right idea. I bought some Boston Baked Beans candy and started to chew on them. I waited for someone to come check their mail and leave through the swinging doors. As they left to go through the doors I went behind him. As he let the door go, I let it hit my foot, but acted like it hit me in the face. I screamed, leaned over and spit out the Boston Baked Beans slowly. It looked just like my teeth were knocked out with the red and white candy. Rick Fern was with me and he was laughing so hard but the guy that 'closed' the door on me didn't think it was funny and I thought he was going to punch me out. I had to quickly apologize or suffer the consequences!

Dude, Where's my car?

This one was sent in anonymously for obvious reasons. One of the first things all the GIs first did when they first arrived at Fitzsimons was to go down on Colfax Avenue to a used car dealer to get their first car. I am sure all the car dealers knew we were easy preys and they were eager to get us to buy from them. My first car in the Army was a 1970 dark brown Ford LTD. A huge car – like most of the cars from the 1970's were. It had four doors, brown cloth interior and a trunk big enough to store a large family. I bought the car and promptly took it home during Thanksgiving break. As expected in

Colorado during November, we had snow and ice all over the highways. Being from the south, I was not used to driving in this type of weather. As I crept down the highway into New Mexico, I kept sliding off the highway into the shoulder and into deeper snow. I finally remembered I had bought snow chains, so I stopped the car and installed the chains. I proudly got back into the car and it wouldn't start. We tried everything but it wouldn't budge. Finally, someone came by and gave us a ride to the next town. A tow truck went back and picked up the car to tow into town. We took a bus and made the trip home. I took a bus back to Aurora from home and went to visit the car salesman to tell him the story. He promptly told me to be quiet and took me aside to tell me my financing did not go through and the car had been reported stolen! He told us we needed to go back there, retrieve the car and bring it back to him. Four of us drove down to New Mexico to retrieve the car but did not know the salesman had not made arrangements for payment of storage of the car! We did not have money, so we left the car there. I wonder what ever happened to that car. I wonder what the statute of limitations is.

Lights, Camera, Action!

When you are going through biomed training at Fitzsimmons you can't wait to graduate and get your photograph on the wall with the other hundreds of classes that have graduated. I wanted to do something special for the photograph so I had a set of Groucho Marx glasses ready to put on just as they took the photo. I guess they took several takes of the class photo since that one did NOT make it to the wall. Lucky for me and you, I have the photo to share.

David Braeutigam and class 4-78 USAMEOUS Aurora, Colorado. David is top right with his Groucho Marx glasses. [56]

Other books by the author

The Elgin J. Luckenbach Story. From Luckenbach, Texas to MIA in New Guinea in WWII to the 62 Year Journey Back Home by David W. Braeutigam 2017

How to Become an HTM Consultant by David W. Braeutigam 2017

Books are available on Amazon.com

References

[1] https://en.wikipedia.org/wiki/Biomedical_Equipment_Technician accessed 29 September 2017

[2] http://accenet.org/about/Documents/What's_a_Clinical_Engineer.pdf accessed 29 September 2017

[3] Photo courtesy of David Braeutigam

[4] ibid

[5] ibid

[6] ibid

[7] No patent but I am Copyrighting the Aqua Fuse name as of this publication – so there!

[8] Photo courtesy of Wayne Shaver

[9] https://www.youtube.com/watch?v=kuHKWzWtBzM accessed 26 September 2017

[10] https://en.wikipedia.org/wiki/Commodore_64 accessed 30 September 2017

[11] https://www.youtube.com/watch?v=60z_1G1N0vc accessed 8 October 2017

[12] Photo courtesy of Daniel Irving

[13] Photo courtesy of David Braeutigam and by doing so breaks the promise to never publish the photo in public. Sorry Binseng the photo is TOO funny!

[14] Photo courtesy of David Braeutigam

[15] https://en.wikipedia.org/wiki/Wite-Out accessed 26 September 2017

[16] https://en.wikipedia.org/wiki/X10_(industry_standard) accessed 26 September 2017

[17] https://en.wikipedia.org/wiki/Commodore_VIC-20 accessed 28 September 2017

[18] Photo courtesy of David Braeutigam

[19] https://en.wikipedia.org/wiki/Wireless_Medical_Telemetry_Service accessed 28 September 2017

[20] https://en.wikipedia.org/wiki/Groucho_Marx accessed 28 September 2017

[21] Photo courtesy of David Braeutigam

[22] https://en.wikipedia.org/wiki/MacGyver accessed 8 Nov 2017

[23] https://en.wikipedia.org/wiki/Mannitol accessed 8 Nov 2017

[24] http://www.urbandictionary.com/define.php?term=wussy accessed 30 September 2017

[25] http://1technation.com/technation-magazine-september-2017/ accessed 9 October 2017

[26] Photo courtesy of Travis Recksiek

[27] Photo courtesy of Doug Dreps

[28] Photo courtesy of Andrea Brainard

[29] Photo courtesy of Jay McLure

[30] Photo courtesy of Cody Brown

[31] Photo courtesy of Mitch Smart
[32] Photo courtesy of Justin Wallace
[33] Photo courtesy of Donny Letson
[34] Photo courtesy of Chace Torres
[35] Photo courtesy of Reginald Jones
[36] Photo courtesy of Chace Torres
[37] Photo courtesy of Justin Donovan
[38] Photo and tagline courtesy of A Zahi Adl
[39] ibid
[40] Photo courtesy of David Mason
[41] Photo courtesy of Jeremy Hendrick
[42] Photo courtesy of Luis Jurado
[43] Photo courtesy of Adam Fiske
[44] Photo courtesy of Matt Burns
[45] Photo courtesy of Mitch Smart
[46] Photo courtesy of Pete Martin
[47] Photo courtesy of A Zahi Adl
[48] Photo and tagline courtesy of Craig Westbrook
[49] ibid
[50] https://en.wikipedia.org/wiki/Honeysucker accessed 26 September 2017
[51] http://www.dhcs.ca.gov/formsandpubs/laws/hipaa/Pages/1.00WhatisHIPAA.aspx accessed 8 October 2017
[52] Photo courtesy of David Braeutigam
[53] https://en.wikipedia.org/wiki/Fitzsimons_Army_Medical_Center accessed 5 Mar 2016
[54] http://www.emsmuseum.org/virtual-museum/Equipment_Manufacturers/articles/398225-1968-Defibrillator-LifePak-33-Serial-0001/ accessed 5 Mar 2016
[55] https://en.wikipedia.org/wiki/Chlorpromazine accessed 28 September 2017
[56] Photo courtesy of David Braeutigam

CPSIA information can be obtained at www.ICGtesting.com
Printed in the USA
LVIW01n1942251217
560734LV00001B/4